Ashish Kumar, Abhinay Thakur (Eds.)
Corrosion Mitigation

Also of Interest

Materials Corrosion and Protection
Huang, Zhang (Eds.), 2018
ISBN 9783110309874, e-ISBN 9783110310054

Active Materials
Fratzl, Friedman, Krauthausen, Schäffner, 2021
ISBN 9783110562064, 9783110561814

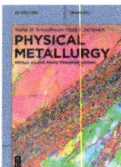

Physical Metallurgy.
Metals, Alloys, Phase Transformations
Schastlivtsev, Zel'dovich, 2022
ISBN 9783110758016, 9783110758023

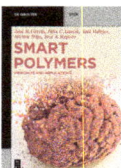

Smart Polymers.
Principles and Applications
García, García, Ruiz, Vallejos, Trigo-López, 2022
ISBN 9781501522406, e-ISBN 9781501522468

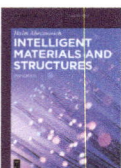

Intelligent Materials and Structures
Abramovich, 2021
ISBN 9783110726695, 9783110726701

Corrosion Mitigation

Biomass and Other Natural Products

Edited by
Ashish Kumar, Abhinay Thakur

DE GRUYTER

Editors
Prof. Ashish Kumar
NCE, Department of Science and Technology,
Government of Bihar
India
drashishchemlpu@gmail.com

Abhinay Thakur
Department of Chemistry
Faculty of Technology and Science
Lovely Professional University
144411 Phagwara, Punjab
India
thakurabhinay96@gmail.com

ISBN 978-3-11-076057-6
e-ISBN (PDF) 978-3-11-076058-3
e-ISBN (EPUB) 978-3-11-076064-4

Library of Congress Control Number: 2022934373

Bibliographic information published by the Deutsche Nationalbibliothek
The Deutsche Nationalbibliothek lists this publication in the Deutsche Nationalbibliografie;
detailed bibliographic data are available on the Internet at http://dnb.dnb.de.

© 2022 Walter de Gruyter GmbH, Berlin/Boston
Cover image: MSchauer/iStock/Getty Images Plus
Typesetting: Integra Software Services Pvt. Ltd.
Printing and binding: CPI books GmbH, Leck

www.degruyter.com

Preface

The phenomenon of corrosion is as old as the history of metals, and it has been looked on as a menace that destroys metals and structures and turns beauty into a beast. Our human civilization cannot exist without metals and yet corrosion is their Achilles' heel. Although the familiarity with corrosion is ancient, it has been taken very passively by scientists and engineers in the past. Surprisingly, it is only during the last six decades that corrosion science has gradually evolved into a well-defined discipline. Corrosion science and engineering is now an integral part of the engineering curriculum in leading universities throughout the world. Due to the worldwide known hazardous effects of toxic chromate corrosion inhibitor being used, researchers are looking for an alternative eco-friendly way of developing nontoxic corrosion inhibitors which cannot be other than the natural sources such as plants, roots, and oils. In this book, all recent advancements in the corrosion field will be discussed with an emphasis on natural sources which is the demand of the era as to replace the commercially available toxic corrosion inhibitors. This book will be found very useful not only for corrosion engineering students but also for corrosion scientists and engineers in solving their problems in their professional capacity and those interested in corrosion. This book gives whole deep insight into details of using the natural sources as corrosion inhibitors and whole essential corrosion measurements and techniques related made in recent years by top researchers and experts working innovatively in the corrosion field, anticipating green earth pledge. In this book, each and every chapter covers the most recent trends in developments and advancements in the field of corrosion where major emphasis is given on the use of natural products such as plant parts, biowaste, and agricultural waste as corrosion inhibitors. The first chapter presents the basic introduction, need for development, and usage of natural corrosion inhibitors which will tally the basic concepts related to it to readers. The second chapter deals with corrosion inhibitors' fundamental concepts and selection metrics which will define the fundamental concepts such as why should we opt for natural inhibitors as corrosion inhibitors while other commercial inhibitors are also in the market. The various selection metrics for delivering potential corrosion inhibition from natural sources will be discussed which is no doubt is a first step toward the efficient success of natural products as corrosion inhibitors. The next chapter deals with the adsorption mechanism of natural corrosion inhibitors as the chemical compounds namely bioactive phytoconstituents protect through adsorption mechanism by chemisorbing at the metal/solution interface on the metal surface forming a protective barrier against corrosion. The paradox is subject to the nature and surface charge of the metal, the nature and type of corrosive solution, and the molecular structure of the inhibiting compound. Chapters 4–6 discuss and enlighten about the usage of various natural products and their sources such as plants, biomass waste, and biopolymers, where the literature time frame will be emphasized from 2010 to 2022, covering innovative research and advancements made by researchers globally in this field. In Chapter 7, we

https://doi.org/10.1515/9783110760583-202

discuss the various methods and techniques frequently used globally by researchers and experts to monitor corrosion and inhibition efficiency exhibited by the inhibitors. In this chapter, several techniques such as EIS, SEM, AFM, gravimetric analysis, and PDP analysis are elaborated efficiently. In Chapter 8, the advantages and disadvantages of using these eco-friendly sustainable natural products as corrosion inhibitors over other commercially available corrosion inhibitors are discussed which will provide an insight idea of using natural inhibitors in this era. After reading it, readers will be able to find out why they should rely upon natural sources among other commercial corrosion inhibitors. In the next chapter, numerous commercialization and economic plus industrial opportunities using these sustainable corrosion inhibitors will be discussed. The global level will be discussed with proper relevant facts and figures which will guide readers to know the positive global impact of using natural products as corrosion inhibitors. Recent patents made in this field will also be discussed. Last but not least, that is, Chapter 10 discusses the challenges and outlooks of using and implementing natural products in the field of corrosion inhibitors. This chapter gives readers an idea about the future research reliability and implementations of these inhibitors globally to a large extent. Several problematic aspects of using natural products as corrosion inhibitors will be explored. Also, the type of challenges faced by researchers or professionals while aiming for the utilization of natural inhibitors on a lab and industrial scale will be explained.

Ashish Kumar, Abhinay Thakur, March 2022

Contents

Muhammad Abubaker Khan, Zahid Nazir, Muhammad Hamza,
Mohammad Tabish, Ghulam Yasin

Humira Assad, Abhinay Thakur, Ayan Bharmal, Shveta Sharma,
Richika Ganjoo, Savas Kaya

Omar Dagdag, Rajesh Haldhar, Seong-Cheol Kim, Elyor Berdimurodov,
Eno E. Ebenso, Savaş Kaya

Ambrish Singh, Kashif Rahmani Ansari, Shivani Singh,
Mumtaz Ahmed Quraishi

Omotayo Sanni, Jianwei Ren, Tien-Chien Jen

Ruby Aslam, Mohammad Mobin, Jeenat Aslam

Richika Ganjoo, Shveta Sharma, Humira Assad, Abhinay Thakur,
Ashish Kumar

List of contributors

Dr. Muhammad Abubaker Khan
College of Physics and Optoelectronic
Engineering
Shenzhen University
Shenzhen 518060, China
abubakarengg@yahoo.com
Chapter 1

Dr. Zahid Nazir
School of Materials Science and Engineering
Beijing Institute of Technology
Beijing 100081, China
Zahid_1623@yahoo.com
Chapter 1

Mr. Muhammad Hamza
Institute of Advance Materials
Bahauddin Zakariya University
Multan, Pakistan
engrhamza@yahoo.com
Chapter 1

Mr. Mohammad Tabish
College of Materials Science and Engineering
Beijing University of Chemical Technology
Beijing 100029, China
tabish.5000@buct.edu.cn
Chapter 1

Dr. Ghulam Yasin
College of Physics and Optoelectronic
Engineering,
Shenzhen University
Shenzhen 518060, China
yasin@szu.edu.cn
Chapter 1

Miss. Humira Assad
Department of Chemistry
Faculty of Technology and Science
Lovely Professional University
Phagwara 144411, Punjab, India
humiraassad888@gmail.com
Chapters 2, 10

Mr. Abhinay Thakur
Department of Chemistry
Faculty of Technology and Science
Lovely Professional University
144411 Phagwara, Punjab, India
thakurabhinay96@gmail.com
Chapters 2, 10

Mr. Ayan Bharmal
Department of Chemistry
Faculty of Technology and Science
Lovely Professional University
Phagwara 144411, Punjab, India
aayansbharmal@gmail.com
Chapter 2

Mrs. Shveta Sharma
Department of Chemistry
Faculty of Technology and Science
Lovely Professional University
Phagwara 144411, Punjab, India
shveta1chem@gmail.com
Chapters 2, 10

Miss. Richika Ganjoo
Department of Chemistry
Faculty of Technology and Science
Lovely Professional University
Phagwara 144411, Punjab, India
ruchikaganjoo3@gmail.com
Chapters 2, 10

Dr. Savas Kaya
Department of Chemistry
Faculty of Science
Cumhuriyet University, Sivas, Turkey
savaskaya1989@gmail.com
Chapter 2

Prof. Ashish Kumar
NCE, Department of Science and Technology,
Government of Bihar, India
drashishchemlpu@gmail.com
Chapters 2, 10

https://doi.org/10.1515/9783110760583-204

Dr. Omar Dagdag
Nanotechnology and Water Sustainability
Research Unit
College of Science, Engineering and
Technology
University of South Africa
Johannesburg 1709, South Africa
omar.dagdag@uit.ac.ma
Chapter 3, 7

Dr. Rajesh Haldhar
School of Chemical Engineering
Yeungnam University
Gyeongsan 712749, South Korea
rajeshhaldhar.lpu@gmail.com
Chapter 3

Dr. Seong-Cheol Kim
School of Chemical Engineering
Yeungnam University
Gyeongsan 712749, South Korea
sckim07@ynu.ac.kr
Chapter 3

Dr. Elyor Berdimurodov
Faculty of Chemistry
National University of Uzbekistan
Tashkent 100034, Uzbekistan
elyor170690@gmail.com
Chapter 3

Prof. Eno E. Ebenso
Nanotechnology and Water Sustainability
Research Unit
College of Science, Engineering and
Technology
University of South Africa
Johannesburg 1709, South Africa
eno.ebenso@gmail.com
Chapter 3

Dr. Savaş Kaya
Department of Pharmacy
Cumhuriyet University Health Services
Vocational School
58140 Sivas, Turkey
savaskaya@cumhuriyet.edu.tr
Chapter 3

Prof. Ambrish Singh
School of New Energy and Materials
Southwest Petroleum University
Chengdu 610500, Sichuan, China
vishisingh4uall@gmail.com
Chapter 4

Dr. Kashif Rahmani Ansari
Center of Research Excellence in Corrosion
Research Institute
King Fahd University of Petroleum and Minerals
Dhahran 31261, Saudi Arabia
ka3787@gmail.com
Chapter 4

Mrs. Shivani Singh
School of Bioengineering and Biosciences
Lovely Professional University
Phagwara 144402, Punjab, India
shivanisinghchina60@gmail.com
Chapter 4

Prof. Mumtaz Ahmed Quraishi
Center of Research Excellence in Corrosion
Research Institute
King Fahd University of Petroleum and Minerals
Dhahran 31261, Saudi Arabia
maquraishi.apc@iitbhu.ac.in
Chapter 4

Dr. Omotayo Sanni
Department of Mechanical Engineering
Science
University of Johannesburg
Cnr Kingsway and University Roads
Auckland Park 2092, Johannesburg
South Africa
tayo.sanni@yahoo.com
Chapter 5

Prof. Jianwei Ren
Department of Mechanical Engineering Science
University of Johannesburg
Cnr Kingsway and University Roads
Auckland Park 2092, Johannesburg
South Africa
jren@uj.ac.za
Chapter 5

Prof. Tien-Chien Jen
Department of Mechanical Engineering
Science
University of Johannesburg
Cnr Kingsway and University Roads
Auckland Park 2092, Johannesburg, South Africa
tjen@uj.ac.za
Chapter 5

Dr. Younes Ahmadi
Department of Analytical Chemistry
Kabul University
Kabul 11001, Afghanistan
and
Department of Civil and Environmental
Engineering
Hanyang University
222 Wangsimni-Ro, Seoul 04763
Republic of Korea
ya021126@gmail.com
Chapter 6

Mr. Mubasher Furmuly
Department of Analytical Chemistry
Kabul University
Kabul 11001, Afghanistan
masih.azizi@gmail.com
Chapter 6

Prof. Nasrin Raji Popalzai
Department of Analytical Chemistry
Kabul University
Kabul 11001, Afghanistan
raji_nasrin@yahoo.com
Chapter 6

Dr. Elyor Berdimurodov
Faculty of Chemistry
National University of Uzbekistan
Tashkent 100034, Uzbekistan
elyor170690@gmail.com
Chapter 7

Prof. Abduvali Kholikov
Faculty of Chemistry
National University of Uzbekistan
Tashkent 100034, Uzbekistan
abduvali0079@gmail.com
Chapter 7

Prof. Khamdam Akbarov
Faculty of Chemistry
National University of Uzbekistan
Tashkent 100034, Uzbekistan
akbarov_Kh@rambler.ru
Chapter 7

Prof. Khasan Berdimuradov
Faculty of Industrial Viticulture and Food
Production Technology
Shahrisabz branch of Tashkent Institute of
Chemical Technology
Shahrisabz 181306, Uzbekistan
khasanberdimuradov@gmail.com
Chapter 7

Dr. Rajesh Haldhar
School of Chemical Engineering
Yeungnam University
Gyeongsan 712749, South Korea
rajeshhaldhar.lpu@gmail.com
Chapter 7

Dr. Mohamed Rbaa
Laboratory of Organic Chemistry Catalysis
and Environment
Faculty of Sciences
Ibn Tofail University
PO Box 133, 14000 Kenitra, Morocco
mohamed.rbaa10@gmail.com
Chapter 7

Dr. Dakeshwar Kumar Verma
Department of Chemistry
Government Digvijay Autonomous
Postgraduate College
Rajnandgaon, Chhattisgarh 491441, India
dakeshwarverma@gmail.com
Chapter 7

Prof. Lei Guo
School of Materials and Chemical Engineering
Tongren University, Tongren 554300, China
and
School of Oil and Natural Gas Engineering
Southwest Petroleum University
Chengdu 610500, China
cqglei@163.com
Chapter 7

Ms. Rajimol Puthenpurackal Ravi
Materials Science and Technology Division
CSIR-National Institute for Interdisciplinary
Science and Technology
Trivandrum 695 019, Kerala, India
and
Academy of Scientific and Innovative
Research (AcSIR)
Ghaziabad 201 002, India
rajipr19@gmail.com
Chapter 8

Dr. Sarah Bill Ulaeto
Materials Science and Technology Division
CSIR-National Institute for Interdisciplinary
Science and Technology
Trivandrum 695 019, Kerala, India
and
Academy of Scientific and Innovative
Research (AcSIR)
Ghaziabad 201 002, India
and
Department of Chemical Sciences
Rhema University Nigeria
Aba, Abia State, Nigeria
sarahbillmails@yahoo.com
Chapter 8

Dr. Thazhavilai Ponnu Devaraj Rajan
Materials Science and Technology Division
CSIR-National Institute for Interdisciplinary
Science and Technology
Trivandrum 695 019, Kerala, India
and
Academy of Scientific and Innovative
Research (AcSIR)
Ghaziabad 201 002, India
tpdrajan@gmail.com
Chapter 8

Dr. Kokkuvayil Vasu Radhakrishnan
Academy of Scientific and Innovative
Research (AcSIR)
Ghaziabad 201 002, India
and
Chemical Science and Technology Division
CSIR-National Institute for Interdisciplinary
Science and Technology
Trivandrum 695 019, Kerala, India
radhu2005@gmail.com
Chapter 8

Dr. Ruby Aslam
Corrosion Research Laboratory
Department of Applied Chemistry
Faculty of Engineering and Technology
Aligarh Muslim University
Aligarh 202002, India
drrubyaslam@gmail.com
Chapter 9

Prof. Mohammad Mobin
Corrosion Research Laboratory
Department of Applied Chemistry
Faculty of Engineering and Technology
Aligarh Muslim University
Aligarh 202002, India
drmmobin@hotmail.com
Chapter 9

Dr. Jeenat Aslam
Department of Chemistry
College of Science
Taibah University
Yanbu 30799
Al-Madina, Saudi Arabia
drjeenataslam@outlook.com
Chapter 9

Muhammad Abubaker Khan, Zahid Nazir, Muhammad Hamza,
Mohammad Tabish, Ghulam Yasin

1 An introduction to natural corrosion inhibitor

Abstract: Chemical components are prone to corrosion due to their everyday use in a variety of human activities/external work environmental conditions. Using inhibitors to prevent chemical components from corrosive environments has become a critical concern. Over several decades, large numbers of organic compounds have been investigated owing to their corrosion inhibition potential. These compounds/chemical inhibitors, particularly those containing nitrogen, silicon, and oxygen species, exhibit significant inhibition activity. Unfortunately, most of these compounds are not only expensive but also harmful to living organisms. Therefore, the importance of natural inhibitors as eco-friendly, readily available, and renewable sources should be explored. The main goal is to highlight the physical and, in some cases, chemical effects of the inhibitors on diverse components in corrosive media. This chapter covers an overview of the adsorption mechanism of natural corrosion inhibitors, inhibitor activity, and the effect of natural inhibitors on metals in corrosive environments.

Keywords: Natural corrosion inhibitors , metals, corrosion, adsorption mechanism, organic compounds

1.1 Introduction

Metallic materials, especially alloys, are extensively used in the building, construction, and petrochemical sectors because of their excellent mechanical strength, corrosion resistance, and low cost. Their use in the petrochemical industry, where most processes like cleaning, pickling, and descaling are all done in an acidic environment, particularly in hydrochloric acid [1–4], poses a challenging environment for metallic applied materials. Corrosion is a significant issue during these processes because of the intense use of acidic solutions that dissolve metallic components and surface contaminants like rusts and scales. Thus, to reduce metallic loss, many external corrosion inhibitors (CIs) are added to the acidic solutions. These supplements absorb and protect metals from harsh conditions [5, 6]. As a result, evaluating the corrosion phenomenon on metals, particularly different types of steel, and developing effective methods are critical. In several situations, different strategies for controlling and preventing corrosion have been developed. For industrial applications, the use of corrosion surface inhibitors is gaining popularity over other protective measures. Many

https://doi.org/10.1515/9783110760583-001

organic inhibitors (OIs) and inorganic inhibitors (IOIs) have been investigated during the past decade to control corrosion of metals [7, 8].

The majority of organic compounds contain nucleophiles such as nitrogen, sulfur, oxygen atoms, heterocyclic compounds, and π-electrons, allowing adsorption on the metal surface [9]. These chemical compounds can adsorb on a metal surface and block the active surface sites to reduce corrosion. However, other organic compounds, including aliphatic amines, aromatic acids, carbonyl compounds, phenol, and amino acids, also show excellent corrosion protection [10–17]. But their toxicity, high cost, and unfriendly nature toward the environment make them unfavorable as CIs [18]. Inorganic chemicals like chromates, dichromates, and nitrates are also used as anodic inhibitors to limit corrosion rates. Unfortunately, the biological toxicity of these inorganic compounds, particularly chromates and organophosphates, indicates that they are environmentally hazardous [8, 19–21]. Although many synthetic chemicals have good CI properties, they are incredibly harmful to humans and the environment. These inhibitors can damage organ systems such as the kidneys or liver, disrupt a biochemical process, or disrupt an enzyme system at a specific location in the body. These harmful effects lead natural inhibitors to grow [9].

Recently, numerous environment-friendly inhibitors have been investigated for applied metallic components over the last several years. In one of the studies, it was portrayed drugs as CIs through an emphasis on efficiency [11, 22]. In two other studies, the importance of biopolymers and surfactants in corrosive conditions is emphasized as well [23, 24]. Another related article discussed using natural ingredients as green corrosion inhibitors (GCIs) in numerous corrosive conditions. Recently, natural inhibitors have emerged as a result of these detrimental influences. Natural inhibitors, such as plant seeds, leaves, lignin, plants extract, and leaf extracts [11, 25–34], contain the essential beneficial compounds for the environment and humans. They have several advantages over chemical inhibitors, including nonpoisonous, inexpensive, renewable, widely available, and environment friendly, making them more applicable [35]. A literature review also revealed that several plant extracts are commonly employed as CIs. Among the extracts tested, leaf extracts had a somewhat higher level of protective efficiency at low concentrations. There are two kinds of extracts: aqueous and organic. Both types of extracts are extensively manufactured and used for various metals and alloys in various electrolytic systems [36, 37].

In light of the necessity of using natural inhibitors in various industries such as the petrochemical industry and others to control corrosion, their desirable features, and the lack of literature on the issue, we have decided to write all related, essential, and comprehensive information about the natural inhibitors and make it accessible to researchers due to the importance of using natural inhibitors in various industries, and to control corrosion, as well as their desirable properties. Figure 1.1 shows the flow diagram for the preparation of plant extract.

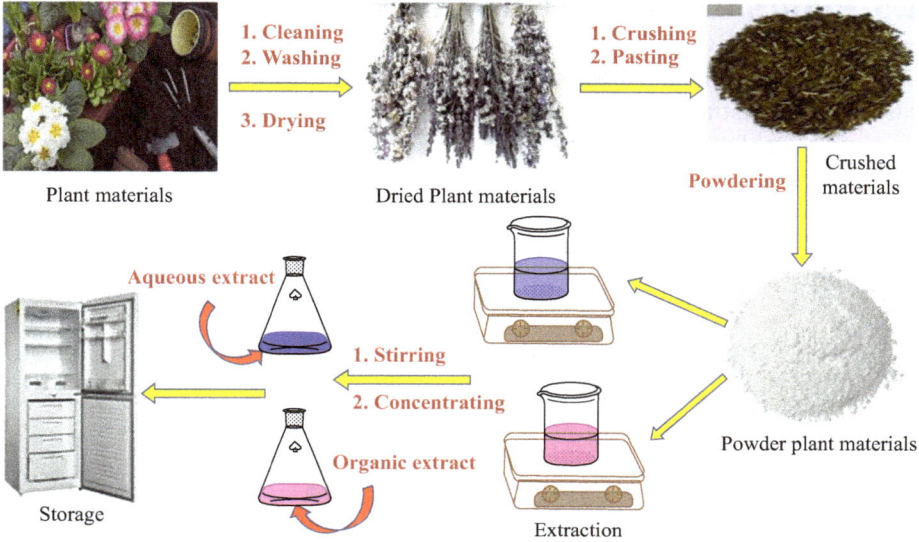

Figure 1.1: Graphic diagram of the preparation of plant extract. Adapted with permission from ref. [38]. Copyright Elsevier (2020).

1.2 Natural corrosion inhibitors

Green inhibitors are widely used in the oil and gas industry to reduce the corrosion of many types of steel in acidic conditions. Green inhibitors are categorized into two types based on their chemical composition: OIs and IOls (Table 1.1). Plants extracts, natural polymers oil extract, ionic liquids, and drugs are organic green inhibitors (OGIs). These compounds offer strong active centers in the form of heteroatoms including high electron density elements like oxygen, nitrogen, and sulfur to be adsorbed at the metal surface. IOIs, for the most part, are harmful and cannot be deliberated as green inhibitors. Inorganic rare-earth elements (salts of lanthanide, for example) have minimal harmfulness and decent biodegradability. OIs are generally more suited to acidic media, while IOIs are better suited to neutral environments. Furthermore, IOIs can be anodic or cathodic, whereas OIs can be both. In general, OGIs outperform inorganic ones in terms of inhibitory efficiency [39–42].

Table 1.1: Natural corrosion inhibitors [43].

OIs	Plants and oils (extract)	Ionic liquids	Amino acid	Natural polymer	Drugs
IOIs			Salt of lanthanide		

1.3 Plants extract as a natural corrosion inhibitor

One of the most common areas of corrosion inhibition research is developing GCIs based on plant extracts. As depicted in Figure 1.2, reports on the use of plant extracts in corrosion inhibition are rising every year. The growing number of articles demonstrates the importance of this topic in finding an excellent elucidation for corrosion harms by applying plant extract-based CIs. The increasing research interest indicates that GCIs derived from plant extracts have a high probability of inhibiting corrosion.

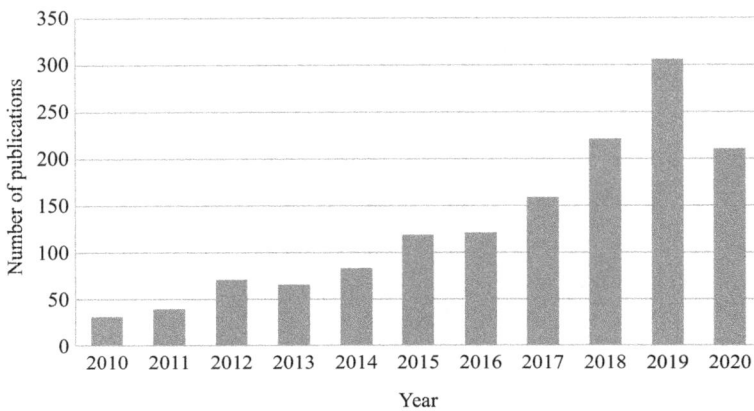

Figure 1.2: The number of articles on GCLs. Adapted with permission from ref. [65]. Copyright Elsevier (2021).

1.3.1 Extraction process

This section focuses on plant extract GCI base. The preparation, inhibition mechanism, and effect of plant extract on corrosion inhibition are briefly explained. In the process of making a plant extract after selecting the plant parts, the extract is dried, ground, and sieved into a powder. Excluding fruit juice extraction, the initial drying phase is required for all plant parts. Room-temperature drying takes time. For example, drying bark takes 20–30 days, either in the shade or sun [44, 45]. After drying, plants can be isolated and extracted using various ways, such as solvent extraction (SE), refinement, pressing, and sublimation [46]. In short, the extraction process involves heating, cooling, and splitting active chemicals in a solvent [47, 48]. Among all of them, SE is the best extensively utilized method for plant extraction. SE requires a solvent to diffuse into plant tissue, solubilize, and extract components (phytochemicals) [49, 50]. Different solvents can be employed to extract different phytochemical concentrations [51]. One researcher recommended that ethanol extract had more tannins total than acetone extract of *Rhizophora*. According to a previous report, most plant extracts are removed with methanol, water, or ethanol [52]. During SE, SE qualities and samples' physical and chemical composition features can be influenced by solvent-to-solid ratio, solvent type, extraction time, and temperature [53]. For example, increasing the solvent-to-solid ratio increased phenol yields from black currants and grape pomace extractions. However, balancing saturation, solvent wastes, and cost impact must be considered. Additionally, smaller particle sizes can also boost extract yield [54, 55].

What's more, the yield of the extract can also be affected by a mixture of organic and aqueous solvents [56, 57]. Because plant extract is soluble in both aqueous and organic solvents, the combined yield is substantially higher than a single system [58]. Mangrove tannins are more soluble in polar than nonpolar solvents. Acetone, a polar solvent, can protect polyphenols from protein interactions, and a mixture of polar solvents can extract more polyphenolic compounds from the bark, such as phenolic content and tannin [59, 60]. Microwave, enzyme, and ultrasound-assisted technologies are also employed for plant extraction [61]. Plant extracts have also been prepared by using compressed liquid extracting agents, such as supercritical fluids, subcritical water, and pressurized fluids. Supercritical fluids are a novel family of substitute solvents for the manufacture of plant extracts that allow for the selective parting of phytochemicals from extracts at modest temperatures and with the shortest treating time [52, 62]. Besides, the temperature during the extraction process is also another issue to consider when preparing plant extracts. The temperature ranging from 60 to 80 °C is preferred for extraction as it is optimal for maximum yield. A suitable extraction temperature is chosen to maximize phytochemical solubility. Higher temperatures increase solubility and mass transfer rates, lowering viscosity and improving solvent flow in matrices. Also, proper extraction temperature is required to avoid the decomposition of active ingredients (phytochemicals). Long extraction times at high temperatures can source the oxidized phenolic chemicals, reducing the yield (extract) [63, 64]. The advantages of these plant extraction technologies are summarized in Table 1.2.

Table 1.2: Plant extraction methods [66–70].

Method	Advantages
SE	A smaller amount of energy required, higher production, fast process, and simplicity of automation
Microwave extraction	Increases reaction yields and decreases reaction times, evades damage, high-temperature heating methods
Enzyme extraction	Increases the overall yield, a new and active method to discharge bounded compounds, permitting plant matrix utilization
Compressed fluids work as removing agents	Permit selective parting of phytochemicals at suitable temperatures and best treating time
Ultrasound extraction	Ultrasound energy facilities organic and inorganic compounds penetrating from the plant matrix

1.3.2 Fruits as natural corrosion inhibitors

Fruits contain various substances, including vitamins, minerals, and phenolic compounds, which can protect metals by adsorbing on their surface and inhibiting active sites for metal dissolution and hydrogen evolution, reducing total metal corrosion in harsh conditions. Many researchers utilized electrochemical impedance measurements, anodic polarization, cathodic curves, and mass loss tests at room temperature to investigate the inhibition effect of fruit extracts [7, 71]. Gomes et al. [72] employed skin extracts of various fruits, including mango, cashew, passion fruit, and orange, as CIs for carbon steel 1020 in 1 M HCl. It was stated that inhibition efficiency (IE) increased as inhibitor concentrations increased and decreased as temperature increased. It was also concluded that fruit peels served as natural inhibitors. In the presence of all extracts, significant inhibition occurred in both cathodic and anodic processes via lowering the current density. The extract of orange peel produced the best IE result (IE: 95% at 400 ppm), while the cashew peel produced the lowest quantity of IE (IE: 80% at 800 ppm) [72]. The concentration of all of the inhibitors mentioned increased polarization resistance. Carotenoids and phenolic compounds from cashew, flavonoids, alkaloids, pectin from passion fruit, flavonoids, carotenoids, pectin from orange, and polyphenols, carotenoids, enzymes, and fiber from mango are the major constituents of inhibitors. One researcher used weight loss, potentiodynamic polarization, and *electrochemical impedance spectroscopy* (EIS) to investigate the anticorrosive impact of *Opuntia elatior* fruit extract on mild steel (MS) in 1 M HCl and H_2SO_4 solution [73]. Another study investigated the effects of apricot juice as a GCI on MS in a 1 M H_3PO_4 solution using the weight loss method at various temperatures [74]. Fruits are another type of natural inhibitor, and they are one of the finest solutions for preserving metals and alloys from corrosion. They are biodegradable and free of heavy metals. These characteristics distinguish them from organic and inorganic banners.

1.3.3 Oils as a natural corrosion inhibitor

One investigator used weight loss, electrochemical polarization, and EIS to explore the inhibitory performance of natural oil extract from pennyroyal mint (*Mentha pulegium*, PM) on the corrosion of steel in 1 M HCl solution [75]. According to the findings, natural oil worked as a cathodic inhibitor by altering the hydrogen reduction pathway. IE was increasing with oil concentration to reach its extreme 80% at 2.76 g/L. It was also boosted by a rise in temperature, indicating chemical adsorption. The adsorption of PM fitted well with Frumkin isotherm. Using polarization curves, one researcher investigated *Thymus satureioids* (TS) oil as a green inhibitor for tinplate in 0.5 M HCl solution. It was discovered that TS operated as a mixed inhibitor with no effect on the mechanism of the hydrogen evolution reaction. The oil's inhibition performance increased with concentration, reaching 87% at 6 g/L, but decreased at high temperatures, reaching 75% at 65 °C [76]. It was discovered that TS was an excellent inhibitor with a physical adsorption mechanism. One other researcher used galvanostatic and potentiodynamic anodic polarization techniques to investigate the inhibitory efficiency of natural black cumin oil on nickel corrosion in 0.1 M HCl solution. It was discovered that IE increased linearly with inhibitor concentration. The adsorption of these compounds followed the Langmuir isotherm [77]. In another related study, one researcher used electrochemical polarization and weight loss measures to investigate the inhibitory impact of *Mentha spicata* essential oil on steel corrosion in 1 M HCl solution. Besides that, Hammuti et al. studied essential oil from fennel (*Foeniculum vulgare*) (FM) as a CI of carbon steel in 1 M HCl utilizing EIS, weight loss measures, and Tafel polarization techniques [78]. Ouachikh et al. also investigated a green corrosion critical oil *Artemisia herba alba* (Art), which, interestingly, reduced the corrosion rate by more than 99.9%. The optimal concentration was 10 ppm. Higher or lower concentrations accelerated corrosion by lowering the surface area covered by the inhibitor and forming unsupported areas on the metal [79]. In fact, the different concentrations were insufficient to establish a protective barrier or the presence of an extra banner encourages the creation of electrostatic repulsion forces between the negative charges, triggering desorption of the banner molecules and the formation of unsupported sites on the metal. Whether in the form of extracts, oils, or pure substances, natural plants may play a significant role in maintaining the environment healthier, safer, and pollution-free. Natural oils' positive results as CIs allowed for the testing of more substance oils.

1.4 Absorption mechanism of natural corrosion inhibitors

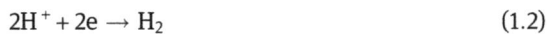

$$Fe \rightarrow Fe^{2+} + 2e \quad (1.1)$$

$$2H^+ + 2e \rightarrow H_2 \quad (1.2)$$

When corrosion inhibitors are introduced into corrosive media, they adsorb on the metal surface's active sites (higher energy areas), forming a protective coating. This layer protects the metal surface against corrosion by isolating it from the corrosive environment. The illustration of the protective coating is depicted schematically in Figure 1.3(a, b). OIs have various advantages over IOIs when it comes to metal surface passivation. For example, Ols can uniformly passivate the surface, providing the best possible protection. But, IOI passive layers are enormously brittle, leaving the metal surface vulnerable to local corrosion attack [80]. Inhibitor adsorption can occur via physical or chemical adsorption or a combination of both (i.e., mixed mode). In any of these two approaches, the surface charge of the substrate affects the interaction of inhibitor molecules with metal surfaces. As shown in Figure 1.3(c1), charged inhibitors interact electrostatically by the active metal surface leading to direct physical adsorption. However, indirect adsorption occurs when preadsorbed halide ions (in amino acids) interact electrostatically with a negatively charged surface, as shown in Figure 1.3(c2).

The metal surface becomes negatively charged due to the adsorption of anions, which increases its ability to adsorb protonated inhibitors. Figure 1.3(c2) shows how this phenomenon can occur in acidic media. In addition, no cation or anion may also be adsorbed on zero-charged metal surfaces (zero point of charge). Therefore, inhibitor adsorption will occur via a chemical reaction between the metal surface and the inhibitor molecules, as shown in Figure 1.3(c3, c4). Inhibitors are electron donors that produce a protective coating by engaging d-orbital of the metal substrate atoms (Figure 1.3(c3)). In some cases, chemical adsorption can occur when metallic ions engage with functional groups (–OH, –NH$_2$, etc.) of green inhibitors, resulting in insoluble complexes that prevent corrosion on the metal surface (Figure 1.3(c4)) [43].

OIs can stop corrosion by forming a protective film on the surface of the metallic material; as previously discussed, IOIs are generally anodic in nature. Their metallic parts are bounded within the film, typically enhancing the corrosion resistance. Various mathematical models, Bockris–Swinkels, Temkin, Flory–Huggins, Frumkin, Freundlich, and Langmuir, have been presented in order to elaborate the extent of adsorption in the absorbent at a fixed temperature [81].

The majority of inhibitors follow the Langmuir isotherm, but some may follow the Frumkin isotherms. The adsorption mechanism is determined by thermodynamic parameters such as the adsorption process's Gibbs free energy ($\Delta G°_{ads}$) entropy and enthalpy. The following equation can be used to calculate the inhibitor's adsorption energy on the steel surface:

$$\Delta G^0_{ads} = - RT \ln(55.5 k_{ads}) \tag{1.3}$$

where R is the gas constant (8.314 J/K mol), T is the absolute temperature in Kelvin, and 55.5 represents the molar concentration value of water molecules in an acidic solution. The relevant adsorption isotherm in Table 1.2 is used to get the value K_{ads}. $\Delta G^0_{ads} \leq -20$ kJ/mol, $\Delta G^0_{ads} \leq -40$ kJ/mol, and 40 kJ/mol $< \Delta G^0_{ads} < -20$ kJ/mol imply physisorption, chemisorption, and mixed adsorption, respectively.

Exothermic adsorption is indicated by enthalpy ΔH^0_{ads} by a negative value. In general, the absolute magnitude ΔH^0_{ads} is higher for chemisorption than for physisorption. To estimate ΔH^0_{ads}, a subsequent version of Langmuir isotherm has been utilized, as follows:

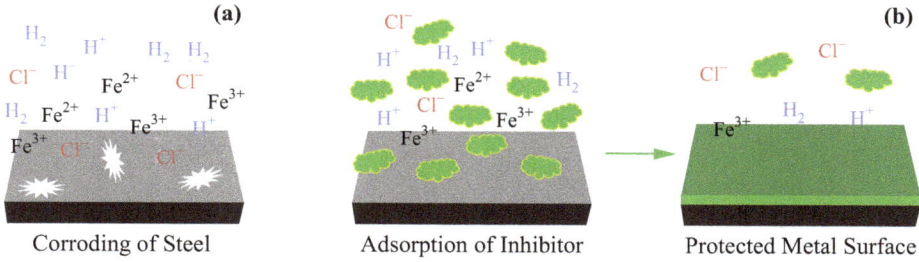

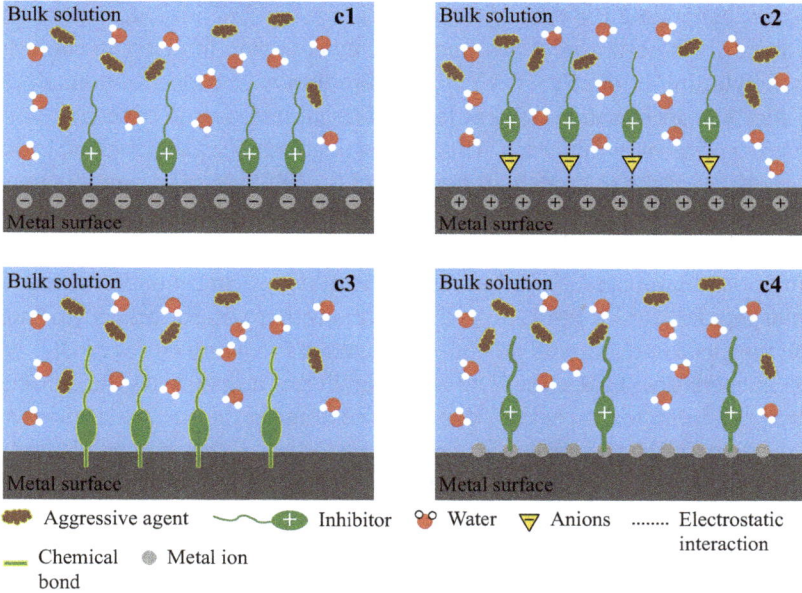

Figure 1.3: (a) Steel corrosion in the absence of inhibitors; (b) corrosion inhibition mechanism; (c1–c4) schematic illustration of natural inhibitors interaction by steel. Adapted with permission from ref. [43]. Copyright Elsevier (2020).

$$\ln k_{ads} = (-\Delta H^0_{ads}/RT) + A \tag{1.4}$$

where ΔH^0_{ads}, T, and A represent the adsorption enthalpy, absolute temperature, and pre-exponential factor, respectively.

The disorder in a system is represented by entropy ΔS^0_{ads}. A value of $\Delta S^0_{ads} < 0$ indicates the existence of the adsorption process. The higher the value the ΔS^0_{ads}, more significant the adsorption will be [82]. The following relation can estimate its value:

$$\Delta G^0_{ads} = \Delta H^0_{ads} - T\Delta S^0_{ads} \tag{1.5}$$

1.5 Factors influencing the efficiency of CIs

The potentiality of an inhibitor to minimize corrosion is denoted as corrosion IE (%). It can be calculated by the corrosion rate and weight loss numbers as follows:

$$C_R = W_{int} - W_{fim}/dAt \tag{1.6}$$

W_{int} and W_{fim} are the metallic samples' original and final weights, respectively. C_R^0 and C_R^i represent the corrosion rates with and without inhibitor molecules, respectively.

The effectiveness of inhibition in preventing corrosion is determined by several parameters, containing the type of the external atmosphere, inhibitor attentiveness, temperature of the solvent and the time spent by corrosion inhibitor in contact with the electrolyte. Inhibitor efficiency usually improves with concentration. However, once a specific concentration is exceeded, additional increases in concentration have little effect on corrosion inhibition. This limit is affected by various factors, including the composition of the electrolyte, the nature of the inhibitor used, and the temperature. Initially, inhibitors are typically deposited flat or horizontally on the metal surface, increasing surface coverage and thus increased inhibition efficacy. However, once an optimal limit is reached, increasing the inhibitor concentration causes intermolecular repulsion. As a result, the inhibitor molecules adsorb vertically or nonplanarly on the metal surface, with no noticeable gain in inhibitory efficacy [83].

The temperature of the electrolyte has a significant influence on the efficiency of inhibition. An inhibitor develops a defensive coating on the surface of the metallic material by chemical adsorption at room temperature. Although chemisorption at higher temperatures is more predominant and has linear relation with the temperature of the surrounding, the growth rate of protective film is also increased which ultimately boosts the inhibition. The contact duration of substrate with the corrosion medium plays equally vital role for the efficiency of natural inhibitors. The relationship between contact time and the effectiveness of the inhibition disclosures important information regarding natural inhibitor stability. These two quantities have an inverse relationship, in general [40, 84, 85]. However, some research has discovered a direct relationship; for instance, the effectiveness of *Clematis gouriana* (a leaf extract) in an acidic solution with a concentration of 400 ppm improves with exposure time [86]. A similar outcome was reported for a natural extract, 2, 3-dihydroxyflavone, as an inhibitor [86].

Another influencing aspect is the structure of an inhibitor which determines the covalent bonding, polarity, steric hindrance and electrostatic effects, and other factors that affect the inhibition efficacy. In general, inhibitors with planar geometry that contain electron-donating groups, for instance (–OH), (–NH$_2$), and (–OCH$_3$) groups, are more efficient than inhibitors with nonplanar geometry that contain extract agents, for example (–NO$_2$) and (–CN) groups. The capacity of natural extracts to suppress corrosion is due to the presence of the aforementioned functional groups in heterocyclic components. Every natural extract has its own organic coordinate, which comprises one or more kinds of electron donor functional groups. The alkaloid pyrrolidine is found in carrots, and the alkaloid ricinine is found in castor seeds. *Lawsonia* extract contains 2-hydroxy-1, 4-naphthoquinone resin, and tannin, as well as coumarin, gallic acid, and sterols. Eucalyptus oil contains monomtrene-1-8-cineole, while *Lawsonia* extract contains 2-hydroxy-1, 4-naphthoquinone resin, and tannin. Hexuronic acid, sugar remains, reducing and nonreducing sugars, and related volatile monoterpenes are all found in the exudate of gums. Primary and secondary amines, unsaturated fatty acids, and bioflavonoids can all be found in *Garcinia kola* seeds. Ascorbic acid, amino acids, flavonoids, pigments, and carotene are among the ingredients in calyx extract [81].

Unusually, several green inhibitors are as effective as hazardous conventional inhibitors; for example, the synthetic inhibitor methyl 3-((2-mercaptophenyl)-imino) butanoate, which works on the chemisorption mechanism, has the maximum efficacy of 99.3%; the triazole derivative 4-amino-5-methyl-4H-1, 2, 4-triazole-3thiol, which works on the physisorption and chemisorption mechanisms, has an efficiency of 99.1%. Black pepper extracts, too, use a mixed-type process and have a 98% efficiency [87]. One researcher tested the aqueous coffee powder extracts as a corrosion inhibition in a 1 M HCl solution [88]. It was reported that as the temperature was raised, the IE became unexpectedly greater than before. The effects of caffeine on carbon steel corrosion in HCl solution have also been investigated [89].

Some scientists investigated the inhibition capability of Schiff base compounds which are synthetic and nontoxic in nature. Due to the electronegative nature of nitrogen, oxygen, and aromatic rings, which are known to have remarkable antiglycating compounds, they act as active adsorption sites on the metallic surface, making them potential green inhibitors [90]. This family of inhibitors has the advantage of ease in production from low-cost precursors. As corrosion inhibitors, numerous potential Schiff base compounds have recently been presented. These can be physically or chemically adsorbed on the steel surface. Due to their high inhibitory efficacy (97%), they can be substituted for hazardous conventional inhibitors [42, 91].

1.5.1 Measuring GCIs efficiency

The first step in determining GCI efficiency is to prepare a metal sample for corrosion testing. Because slight changes in metal composition or accessible impurities during manufacturing affect the results, choosing metal coupons for GCI testing is critical. The metal composition should be as close to corrosion-prone metals as feasible. Moreover, weight loss, electrochemical impedance spectroscopy, linear polarization resistance, and potentiodynamic polarization are popular methods for assessing GCI efficiency.

1.6 Industrial applications of GCIs

CIs derived from natural green materials have industrial applications in petroleum production, steel pipeline manufacturing, refrigerating industry, vehicle, paint industry, boiler, water transmission industry, building construction industry, acid-producing enterprises, and so on.

1.7 Conclusion

Corrosion in metallic materials is a natural phenomenon, but it is dangerous to humans and the atmosphere. The best applied solution is to use CIs as a protective measure to slow down the corrosion. As an alternative to conventional CIs, the use of natural inhibitors as corrosion prevention has been extensively reported in the literature. Numerous plant and fruit parts can be extracted, and the results show that extracts have a high IE in metal corrosion. The presence of heteroatoms and π-electrons in plant extracts affects the IE of GCIs. GCIs, whether in the form of extracts, oils, or pure molecules, may play a significant role in maintaining the environment healthy, safe, and pollution free. Some practical and theoretical evaluation approaches have validated the remarkable efficiency of plant and fruit extracts as GCIs. Though in order to achieve a complete green synthesis of GCIs, several concerns must be addressed, and the use or manufacturing of any harmful solution must be minimized. Finally, the compilation and issues that impact GCI characteristics and efficiency are intended to support other investigators to understand better the significance of GCI evaluation in metal corrosion prevention.

References

[1] Anjum MJ, Ali H, Khan WQ, Zhao J, Yasin G. Metal/metal oxide nanoparticles as corrosion inhibitors. Corrosion Protection at the Nanoscale, 2020; 1: 181–201.
[2] Yasin G, Arif M, Mehtab T, Shakeel M, Khan MA, Khan WQ. Metallic nanocomposite coatings. Corrosion Protection at the Nanoscale, 2020; 1: 245–274.
[3] Hussain MI, Nawaz S, Sajid MM, Nawaz A, Irum A, Javed Y, Ge C, Yasin G. Chapter 5 – Corrosion resistance of nanostructured metals and alloys. *Corrosion Protection at the Nanoscale*. Elsevier. 2020; 63–87.
[4] Yasin G, Anjum MJ, Malik MU, Khan MA, Khan WQ, Arif M, Mehtab T, Nguyen TA, Slimani Y, Tabish M, Ali D, Zuo Y. Revealing the erosion-corrosion performance of sphere-shaped morphology of nickel matrix nanocomposite strengthened with reduced graphene oxide nanoplatelets. Diamond and Related Materials, 2020; 104: 107763.
[5] Fayomi O, Abdulwahab M, Popoola A. Electro-oxidation behaviour and passivation potential of natural oil as corrosion inhibitor in hydrochloric acid environment. International Journal of Electrochemical Science, 2013; 8: 12088–12096.

[6]   Soltani N, Khayatkashani M. Gundelia tournefortii as a green corrosion inhibitor for mild steel in HCl and H2SO4 solutions. International Journal of Electrochemical Science, 2015; 10: 46–62.

[7]   Da Rocha JC, Gomes JADCP, D'Elia E. Corrosion inhibition of carbon steel in hydrochloric acid solution by fruit peel aqueous extracts. Corrosion Science, 2010; 52: 2341–2348.

[8]   Soltani N, Tavakkoli N, Kashani MK, Mosavizadeh A, Oguzie E, Jalali M. Silybum marianum extract as a natural source inhibitor for 304 stainless steel corrosion in 1.0 M HCl. Journal of Industrial and Engineering Chemistry, 2014; 20: 3217–3227.

[9]   Ostovari A, Hoseinieh S, Peikari M, Shadizadeh S, Hashemi S. Corrosion inhibition of mild steel in 1 M HCl solution by henna extract: A comparative study of the inhibition by henna and its constituents (lawsone, gallic acid, α-d-glucose and tannic acid). Corrosion Science, 2009; 51: 1935–1949.

[10] Desai M, Thakar B, Chhaya P, Gandhi M. Inhibition of corrosion of aluminium-51S in hydrochloric acid solutions. Corrosion Science, 1976; 16: 9–24.

[11] Obot I, Obi-Egbedi N, Umoren S. The synergistic inhibitive effect and some quantum chemical parameters of 2, 3-diaminonaphthalene and iodide ions on the hydrochloric acid corrosion of aluminium. Corrosion Science, 2009; 51: 276–282.

[12] Hassan S, Moussa M, El-Tagoury M, Radi A. Aromatic acid derivatives as corrosion inhibitors for aluminium in acidic and alkaline solutions. Anti-Corrosion Methods and Materials, 1990; 37: 8–11.

[13] Issa I, Moussa M, Ghandour M. A study on the effect of some carbonyl compounds on the corrosion of aluminium in hydrochloric acid solution. Corrosion Science, 1973; 13: 791–797.

[14] Lashgari M, Malek AM. Fundamental studies of aluminum corrosion in acidic and basic environments: Theoretical predictions and experimental observations. Electrochimica Acta, 2010; 55: 5253–5257.

[15] Moussa M, Taha F, Gouda M, Singab G. The effect of some hydrazine derivatives on the corrosion of Al in HCl solution. Corrosion Science, 1976; 16: 379–385.

[16] Hassan SM, Elawady Y, Ahmed AI, Baghlaf A. Studies on the inhibition of aluminium dissolution by some hydrazine derivatives. Corrosion Science, 1979; 19: 951–959.

[17] Bereket G, Yurt A. The inhibition effect of amino acids and hydroxy carboxylic acids on pitting corrosion of aluminum alloy 7075. Corrosion Science, 2001; 43: 1179–1195.

[18] Deng S, Li X. Inhibition by Jasminum nudiflorum Lindl. leaves extract of the corrosion of aluminium in HCl solution. Corrosion Science, 2012; 64: 253–262.

[19] Ilevbare G, Burstein G. The inhibition of pitting corrosion of stainless steels by chromate and molybdate ions. Corrosion Science, 2003; 45: 1545–1569.

[20] Badawy W, Al-Kharafi F. The inhibition of the corrosion of Al, Al-6061 and Al-Cu in chloride free aqueous media: I. Passivation in acid solutions. Corrosion Science, 1997; 39: 681–700.

[21] Wang H, Akid R. Encapsulated cerium nitrate inhibitors to provide high-performance anti-corrosion sol–gel coatings on mild steel. Corrosion Science, 2008; 50: 1142–1148.

[22] El-Dahan H, Soror T, El-Sherif R. Studies on the inhibition of aluminum dissolution by hexamine–halide blends: Part I. Weight loss, open circuit potential and polarization measurements. Materials Chemistry and Physics, 2005; 89: 260–267.

[23] Sushmitha Y, Rao P. Material conservation and surface coating enhancement with starch-pectin biopolymer blend: A way towards green. Surfaces and Interfaces, 2019; 16: 67–75.

[24] Shahini MH, Ramezanzadeh B, Mohammadloo HE. Recent advances in biopolymers/carbohydrate polymers as effective corrosion inhibitive macro-molecules: A review study from experimental and theoretical views. Journal of Molecular Liquids, 2021; 325: 115–110.

[25] Zucchi F, Omar IH. Plant extracts as corrosion inhibitors of mild steel in HCl solutions. Surface Technology, 1985; 24: 391–399.

[26] Deng S, Li X. Inhibition by Ginkgo leaves extract of the corrosion of steel in HCl and H2SO4 solutions. Corrosion Science, 2012; 55: 407–415.

[27] Abiola OK, James A. The effects of Aloe vera extract on corrosion and kinetics of corrosion process of zinc in HCl solution. Corrosion Science, 2010; 52: 661–664.

[28] Farooqi I, Quraishi M, Saini P. Corrosion prevention of mild steel in 3% NaCl water by some naturally-occurring substances. Corrosion Prevention and Control, 1999; 70: 370–372.

[29] Abiola OK, Otaigbe J, Kio O. Gossipium hirsutum L. extracts as green corrosion inhibitor for aluminum in NaOH solution. Corrosion Science, 2009; 51: 1879–1881.

[30] Abdel-Gaber A, Abd-El-Nabey B, Sidahmed I, El-Zayady A, Saadawy M. Inhibitive action of some plant extracts on the corrosion of steel in acidic media. Corrosion Science, 2006; 48: 2765–2779.

[31] Valek L, Martinez S. Copper corrosion inhibition by Azadirachta indica leaves extract in 0.5 M sulphuric acid. Materials Letters, 2007; 61: 148–151.

[32] Li X, Deng S, Fu H. Inhibition of the corrosion of steel in HCl, H2SO4 solutions by bamboo leaf extract. Corrosion Science, 2012; 62: 163–175.

[33] Ren Y, Luo Y, Zhang K, Zhu G, Tan X. Lignin terpolymer for corrosion inhibition of mild steel in 10% hydrochloric acid medium. Corrosion Science, 2008; 50: 3147–3153.

[34] Chauhan L, Gunasekaran G. Corrosion inhibition of mild steel by plant extract in dilute HCl medium. Corrosion Science, 2007; 49: 1143–1161.

[35] Fares MM, Maayta A, Al-Mustafa JA. Corrosion inhibition of iota-carrageenan natural polymer on aluminum in presence of zwitterion mediator in HCl media. Corrosion Science, 2012; 65: 223–230.

[36] Baly ECC, Heilbron IM, Barker WF. CX. – Photocatalysis. Part I. The synthesis of formaldehyde and carbohydrates from carbon dioxide and water. Journal of the Chemical Society, Transactions, 1921; 119: 1025–1035.

[37] Schreiner M, Huyskens-Keil S. Phytochemicals in fruit and vegetables: Health promotion and postharvest elicitors. Critical Reviews in Plant Sciences, 2006; 25: 267–278.

[38] Alrefaee SH, Rhee KY, Verma C, Quraishi M, Ebenso EE. Challenges and advantages of using plant extract as inhibitors in modern corrosion inhibition systems: Recent advancements. Journal of Molecular Liquids, 2020; 321: 114666.

[39] Verma C, Ebenso EE, Quraishi M. Ionic liquids as green and sustainable corrosion inhibitors for metals and alloys: An overview. Journal of Molecular Liquids, 2017; 233: 403–414.

[40] Nair RN. Green Corrosion Inhibitors for mild steel in acidic medium. International Journal of Modern Trends in Engineering and Research, 2017; 4: 216–221.

[41] Popoola LT. Progress on pharmaceutical drugs, plant extracts and ionic liquids as corrosion inhibitors. Heliyon, 2019; 5: 1143.

[42] Saxena N, Kumar S, Sharma M, Mathur S. Corrosion inhibition of mild steel in nitric acid media by some Schiff bases derived from anisalidine. Polish Journal of Chemical Technology, 2013; 15: 61–67.

[43] Wei H, Heidarshenas B, Zhou L, Hussain G, Li Q, Ostrikov KK. Green inhibitors for steel corrosion in acidic environment: State of art. Materials Today Sustainability, 2020; 10: 100044.

[44] Marsoul A, Ijjaali M, Elhajjaji F, Taleb M, Salim R, Boukir A. Phytochemical screening, total phenolic and flavonoid methanolic extract of pomegranate bark (Punica granatum L): Evaluation of the inhibitory effect in acidic medium 1 M HCl. Materials Today: Proceedings, 2020; 27: 3193–3198.

[45] Marzorati S, Verotta L, Trasatti SP. Green corrosion inhibitors from natural sources and biomass wastes. Molecules, 2019; 24: 48.

[46] Meriem-Benziane M, Bou-Saïd B, Boudouani N. The effect of crude oil in the pipeline corrosion by the naphthenic acid and the sulfur: A numerical approach. Journal of Petroleum Science and Engineering, 2017; 158: 672–679.

[47] Miralrio A, Espinoza Vázquez A. Plant extracts as green corrosion inhibitors for different metal surfaces and corrosive media: A review. Processes, 2020; 8: 942.

[48] Miri R, Chouikhi A. Ecotoxicological marine impacts from seawater desalination plants. Desalination, 2005; 182: 403–410.

[49] Mobin M, Basik M, Shoeb M. A novel organic-inorganic hybrid complex based on Cissus quadrangularis plant extract and zirconium acetate as a green inhibitor for mild steel in 1 M HCl solution. Applied Surface Science, 2019; 469: 387–403.

[50] Mourya P, Banerjee S, Singh M. Corrosion inhibition of mild steel in acidic solution by Tagetes erecta (Marigold flower) extract as a green inhibitor. Corrosion Science, 2014; 85: 352–363.

[51] Muhamad II, Hassan ND, Mamat SN, Nawi NM, Rashid WA, Tan NA. Extraction technologies and solvents of phytocompounds from plant materials: Physicochemical characterization and identification of ingredients and bioactive compounds from plant extract using various instrumentations. *Ingredients Extraction by Physicochemical Methods in Food*. 2017, 523–560. Academic Press.

[52] El Ibrahimi B, Jmiai A, Bazzi L, El Issami S. Amino acids and their derivatives as corrosion inhibitors for metals and alloys. Arabian Journal of Chemistry, 2020; 13: 740–771.

[53] Dai J, Mumper RJ. Plant phenolics: Extraction, analysis and their antioxidant and anticancer properties. Molecules, 2010; 15: 7313–7352.

[54] Dargahi M, Olsson A, Tufenkji N, Gaudreault R. Green technology: Tannin-based corrosion inhibitor for protection of mild steel. Corrosion, 2015; 71: 1321–1329.

[55] Dehghani A, Bahlakeh G, Ramezanzadeh B. A detailed electrochemical/theoretical exploration of the aqueous Chinese gooseberry fruit shell extract as a green and cheap corrosion inhibitor for mild steel in acidic solution. Journal of Molecular Liquids, 2019; 282: 366–384.

[56] Dehghani A, Bahlakeh G, Ramezanzadeh B, Ramezanzadeh M. Applying detailed molecular/ atomic level simulation studies and electrochemical explorations of the green inhibiting molecules adsorption at the interface of the acid solution-steel substrate. Journal of Molecular Liquids, 2020; 299: 112220.

[57] Dehghani A, Bahlakeh G, Ramezanzadeh B, Ramezanzadeh M. Potential of Borage flower aqueous extract as an environmentally sustainable corrosion inhibitor for acid corrosion of mild steel: Electrochemical and theoretical studies. Journal of Molecular Liquids, 2019; 277: 895–911.

[58] Divya P, Subhashini S, Prithiba A, Rajalakshmi R. Tithonia diversifolia flower extract as green corrosion inhibitor for mild steel in acid medium. Materials Today: Proceedings, 2019; 18: 1581–1591.

[59] Dranca F, Oroian M. Extraction, purification and characterization of pectin from alternative sources with potential technological applications. Food Research International, 2018; 113: 327–350.

[60] Dugo P, Mondello L, Errante G, Zappia G, Dugo G. Identification of anthocyanins in berries by narrow-bore high-performance liquid chromatography with electrospray ionization detection. Journal of Agricultural and Food Chemistry, 2001; 49: 3987–3992.

[61] Eddy NO, Odoemelam SA, Odiongenyi AO. Inhibitive, adsorption and synergistic studies on ethanol extract of Gnetum africana as green corrosion inhibitor for mild steel in H2SO4. Green Chemistry Letters and Reviews, 2009; 2: 111–119.

[62] El-Yaktini A, Lachiri A, El-Faydy M, Benhiba F, Zarrok H, El-Azzouzi M, Zertoubi M, Azzi M, Lakhrissi B, Zarrouk A. Practical and theoretical study on the inhibitory influences of new Azomethine derivatives containing an 8-hydroxyquinoline moiety for the corrosion of carbon steel in 1 M HCl. Oriental Journal of Chemistry, 2018; 34: 3016.

[63] Faisal M, Saeed A, Shahzad D, Abbas N, Larik FA, Channar PA, Fattah TA, Khan DM, Shehzadi SA. General properties and comparison of the corrosion inhibition efficiencies of the triazole derivatives for mild steel. Corrosion Reviews, 2018; 36: 507–545.

[64] Faiz M, Zahari A, Awang K, Hussin H. Corrosion inhibition on mild steel in 1 M HCl solution by Cryptocarya nigra extracts and three of its constituents (alkaloids). RSC Advances, 2020; 10: 6547–6562.

[65] Salleh SZ, Yusoff AH, Zakaria SK, Taib MAA, Seman AA, Masri MN, Mohamad M, Mamat S, Sobri SA, Ali A. Plant extracts as green corrosion inhibitor for ferrous metal alloys: A review. Journal of Cleaner Production, 2021; 304: 127030.

[66] Chen H, Wang L. Chapter 3 – Pretreatment strategies for biochemical conversion of biomass. *Technologies for Biochemical Conversion of Biomass*. 2017, 21–64. Elsevier.

[67] Jin R, Fan L, An X. Microwave assisted ionic liquid pretreatment of medicinal plants for fast solvent extraction of active ingredients. Separation and Purification Technology, 2011; 83: 45–49.

[68] Rosenthal A, Pyle D, Niranjan K. Aqueous and enzymatic processes for edible oil extraction. Enzyme and Microbial Technology, 1996; 19: 402–420.

[69] Herrera M, De Castro ML. Ultrasound-assisted extraction for the analysis of phenolic compounds in strawberries. Analytical and Bioanalytical Chemistry, 2004; 379: 1106–1112.

[70] Verma C, Ebenso EE, Bahadur I, Quraishi M. An overview on plant extracts as environmental sustainable and green corrosion inhibitors for metals and alloys in aggressive corrosive media. Journal of Molecular Liquids, 2018; 266: 577–590.

[71] Flores-De los Ríos J, Sánchez-Carrillo M, Nava-Dino C, Chacón-Nava J, González-Rodríguez J, Huape-Padilla E, Neri-Flores M, Martínez-Villafañe A. Opuntia ficus-indica extract as green corrosion inhibitor for carbon steel in 1 M HCl solution. Journal of Spectroscopy, 2015; 714692: 9.

[72] Gomes JADCP, Rocha JC, D'elia E. Use of fruit skin extracts as corrosion inhibitors and process for producing same. Google Patents, 2015; 8: 926–867.

[73] Loganayagi C, Kamal C, Sethuraman M. Opuntiol: An active principle of Opuntia elatior as an eco-friendly inhibitor of corrosion of mild steel in acid medium. ACS Sustainable Chemistry & Engineering, 2014; 2: 606–613.

[74] Yaro AS, Khadom AA, Wael RK. Apricot juice as green corrosion inhibitor of mild steel in phosphoric acid. Alexandria Engineering Journal, 2013; 52: 129–135.

[75] Bouyanzer A, Hammouti B, Majidi L. Pennyroyal oil from Mentha pulegium as corrosion inhibitor for steel in 1 M HCl. Materials Letters, 2006; 60: 2840–2843.

[76] Bammou L, Chebli B, Salghi R, Bazzi L, Hammouti B, Mihit M, Idrissi H. Thermodynamic properties of Thymus satureioides essential oils as corrosion inhibitor of tinplate in 0.5 M HCl: Chemical characterization and electrochemical study. Green Chemistry Letters and Reviews, 2010; 3: 173–178.

[77] Abdallah M, Al Karanee S, Abdel Fatah A. Inhibition of acidic and pitting corrosion of nickel using natural black cumin oil. Chemical Engineering Communications, 2010; 197: 1446–1454.

[78] Znini M, Bouklah M, Majidi L, Kharchouf S, Aouniti A, Bouyanzer A, Hammouti B, Costa J, Al-Deyab S. Chemical composition and inhibitory effect of Mentha spicata essential oil on the corrosion of steel in molar hydrochloric acid, International Journal of Electrochemical Science, 2011; 6: 691–704.

[79] Ouachikh O, Bouyanzer A, Bouklah M, Desjobert J-M, Costa J, Hammouti B, Majidi L. Application of essential oil of Artemisia herba alba as green corrosion inhibitor for steel in 0.5 MH 2 SO 4. Surface Review and Letters, 2009; 16: 49–54.

[80] Ali SA, El-Shareef A, Al-Ghamdi R, Saeed M. The isoxazolidines: The effects of steric factor and hydrophobic chain length on the corrosion inhibition of mild steel in acidic medium. Corrosion Science, 2005; 47: 2659–2678.

[81] Popoola LT. Organic green corrosion inhibitors (OGCIs): A critical review. Corrosion Reviews, 2019; 37: 71–102.

[82] Fawzy A, Abdallah M, Zaafarany I, Ahmed S, Althagafi I. Thermodynamic, kinetic and mechanistic approach to the corrosion inhibition of carbon steel by new synthesized amino acids-based surfactants as green inhibitors in neutral and alkaline aqueous media. Journal of Molecular Liquids, 2018; 265: 276–291.

[83] Cen H, Chen Z, Guo X. N, S co-doped carbon dots as effective corrosion inhibitor for carbon steel in CO2-saturated 3.5% NaCl solution. Journal of the Taiwan Institute of Chemical Engineers, 2019; 99: 224–238.

[84] Verma C, Haque J, Quraishi M, Ebenso EE. Aqueous phase environmental friendly organic corrosion inhibitors derived from one step multicomponent reactions: A review. Journal of Molecular Liquids, 2019; 275: 18–40.

[85] Rathi P, Trikha S, Kumar S. Plant extracts as green corrosion inhibitors in various corrosive media-a review. World Journal of Pharmacy and Pharmaceutical Sciences, 2017; 6: 482–514.

[86] Martinez S, Stern I. Thermodynamic characterization of metal dissolution and inhibitor adsorption processes in the low carbon steel/mimosa tannin/sulfuric acid system. Applied Surface Science, 2002; 199: 83–89.

[87] Raja PB, Sethuraman MG. Inhibitive effect of black pepper extract on the sulphuric acid corrosion of mild steel. Materials Letters, 2008; 62: 2977–2979.

[88] Torres VV, Amado RS, De Sá CF, Fernandez TL, da Silva Riehl CA, Torres AG, D'Elia E. Inhibitory action of aqueous coffee ground extracts on the corrosion of carbon steel in HCl solution. Corrosion Science, 2011; 53: 2385–2392.

[89] Awad MI. Eco friendly corrosion inhibitors: Inhibitive action of quinine for corrosion of low carbon steel in 1 M HCl. Journal of Applied Electrochemistry, 2006; 36: 1163–1168.

[90] Singh P, Chauhan D, Chauhan S, Singh G, Quraishi M. Chemically modified expired Dapsone drug as environmentally benign corrosion inhibitor for mild steel in sulphuric acid useful for industrial pickling process. Journal of Molecular Liquids, 2019; 286: 110903.

[91] Verma CB, Quraishi M. Schiff's bases of glutamic acid and aldehydes as green corrosion inhibitor for mild steel: Weight-loss, electrochemical and surface analysis. International Journal of Innovative Science Engineering and Technology, 2014; 3: 14601–14613.

Humira Assad, Abhinay Thakur, Ayan Bharmal, Shveta Sharma,
Richika Ganjoo, Savas Kaya

2 Corrosion inhibitors: fundamental concepts and selection metrics

Abstract: Corrosion is the gradual breakdown of substances (typically metals) determined by the chemical and/or electrochemical processes with their environment over time. Corrosion has become a critical concern since it not only damages metals and alloys but also impacts our supplies and the menial work engaged in the production of various metallic instruments and commodities. Metals and alloys used in a wide range of sectors are just more prone to corrosion, which cannot be avoided but can be dealt with in a multitude of approaches. Among the different strategies for avoiding or preventing metal surface deterioration or depreciation, the application of corrosion inhibitors (CIs) is among the greatest competent strategies for shielding metal surfaces against corrosion. Owing to its low price and simplicity of use, this approach is rapidly gaining popularity. In the past, inhibitors were widely used in industries due to their outstanding anticorrosive characteristics. Their efficiency is determined by their chemical makeup, crystal structures, and adsorption strengths on the surface of the metal. This chapter provides an overview of the function of CIs, advancement in the development of CIs, and CI eligibility requirements based on the method of shielding, conditions, underlying mechanisms, and so on. Furthermore, this chapter exhibits CI applications and explores simply the good advancements and patents in the use of CIs for metallic substances in various situations.

Keywords: Corrosion, corrosion inhibitors, adsorption, inhibition mechanism

2.1 Introduction

Corrosion is the degradation or disintegration of metallic substances and alloys in the influence of an atmosphere through chemical or electrochemical causes [1]. Corrosive or abrasive medium refers to the environment in which the metal corrodes. Steel, domestic, and gardening equipment rusting outdoors is a typical occurrence. Corrosion products are chemical molecules that contain the metal in its oxidized form. Apart from gold (Au) and platinum (Pt), all other metallic elements deteriorate and convert themselves into compounds comparable to the primary minerals from which they are mined [2]. Corrosion, on the other hand, is equally as widespread in other material groups including ceramics, plastics, and rubber. Meanwhile, almost all surroundings are destructive to a specific extent, they are important factors/subsidizing drivers of progressive collapse and impose a significant economic

https://doi.org/10.1515/9783110760583-002

burden on society [3]. Corrosion, in addition to causing catastrophic failure, harms not just the surroundings but also human health and safety and industrial plants. Comprehension of corrosion and the implementation of prompt and effective preventive actions are critical in the prevention of corrosion breakdowns [4]. Plating, alloying, cathodic and anodic protection, and subsequently been employing the laser for this goal by interface modification of metallic substance is viewed as a method to enhance the qualities of materials such as toughness, stiffness, and corrosion resistance [5, 6]. Corrosion inhibitors (CI), on the other hand, are of substantial practical value, since they are widely used in decreasing metallic scrap throughout the manufacturing and minimizing the danger of metal fatigue, both of which can result in the abrupt cessation of manufacturing processes, resulting in substantial expenditures. It is also critical to apply CIs to avoid mineral breakdown and limit the usage of acidic chemicals [7, 8]. Owing to its low cost and ease of use, CIs are widely used in repressing or at least minimizing metal dissolution in a variety of disciplines, ranging from manufacturing industries to construction equipment to surface remedies for cultural assets. Chemical inhibitors are used in a variety of ways to slow down degradation mechanisms. In the oil harvesting and manufacturing sectors, CIs have historically been regarded as the first line of protection. A CI is defined as a "chemical compound that, when introduced in the corrosive environment at an appropriate quantity, reduces the rate of corrosion without materially affecting the saturation of any corrosive agent [9]." In general, it is efficient in low quantities. This includes any agent that slows the corrosion process by creating significant pH fluctuation, as well as oxygen and hydrogen sulfides scavengers that cause hostile species to be removed from the environment. Inhibitors are used in a variety of industrial and corporate services and devices, including air conditioning systems, processing plants, pipelines, chemicals, oil and gas generation regiments, furnaces and water purification, acrylics, dyes, lubricant materials, and so on [10]. There is documentation of the use of inhibitors dating back to the early nineteenth century. They were already being utilized to safeguard metals in procedures like acid harvesting, shielding against hostile water, acidified oilfields, and cooling equipment at the age. Since the 1950s and 1960s, there have been considerable developments in the advancement of CI technologies, such as the use of electrochemical processes to investigate CIs [11]. According to the recent estimates, the requirement for CIs (such as organic, inorganic, alkaline, and neutral) in the United States would increase by 4.1% annually to US 25×10^{-1} billion in 2017. In 2012, they predicted that 26.6% of the customer preferences for CIs was for processing gasoline, 16.9% for practicalities, 16.7% for petroleum generation, 15.3% for chemical, 9.5% for metallic substances, 7.1% for pulp and paper components, and 8.0% for others [12]. A significant number of publications have been documented in order to produce ecologically suitable CIs, and an extensive study has been undertaken to produce more efficient and less toxic CIs [13]. In addition, there has been an increase in the study of naturally occurring products like plant excerpts, essential oils, and refined components to obtain environmentally approachable compounds. A large number of technical investigations have been conducted on CIs. However, the majority of what is recognized is the result of trial-and-error investigations, both in institutions and in the outdoors. There are few rules, equations, or hypotheses to govern inhibitor advancement and use. This chapter attempts to analyze a variety of CIs stemming from different criteria and to investigate their inhibition efficiency in diverse corrosive fluids. Furthermore, advancements in the domain of CIs and their applicability for the corrosion protection of metals in many areas have also been discussed.

2.2 Chemistry and adverse impact of corrosion

Metals, in general, have distinctive qualities such as being opaque, glossy, conductive, malleable, and ductile, and they enthusiastically produce ionic bonds with nonmetals and metal connections with other metallic materials [14]. Metals with electrical structures include overlapping conduction and valence bands. Metals are extracted from their ores by expending a great deal of energy. Heat is stored as potential energy in metal owing to the processing and purifying processes, and this sort of energy is released throughout the electrochemical reaction after interacting with the surroundings. These metals are thus in thermodynamic equilibrium and will end up losing energy by retreating to molecules that are chemically close to their initial forms; for instance, the core constituent for Fe and steel production and the corrosion moiety "rust" (Fe_2O_3) have similar chemical makeup. The energy accumulated during melting and liberated during dissolution acts as a driving force for the corrosion reaction to continue. The Indian government spends approximately 3.5% of the nation's gross domestic product each year on corrosion losses [15]. According to recent studies, not only India but also other countries are increasing their expenditures to meet the demand for CIs. Because most metallurgical substances, particularly corrosion compounds, have low mechanical properties, a badly corroded metal plate is rendered completely worthless for its intended purpose. Metals that demand extra energy to refine, like Mg, Al, Zn, and Fe, are far more prone to corrosion than materials that need lower energy to reprocess like Au, Ag, and Pt. A corrosion cycle is shown in Figure 2.1.

Thus, deterioration has a far greater impact on the relatively secure, dependable, and convenient operation of the machine and constructions than simply losing the density of the material. Even though the volume of material affected is minimal, malfunctions of various machines and the need for costly restorations are possible. The following are some of the most serious negative impacts of corrosion:

- Nuclear power plants are forced to close owing to failures, such as a nuclear reactor failing during the purification procedure that may lead to several problems to commercial sectors and consumers.
- Expensive hardware overhaul due to rusted parts.

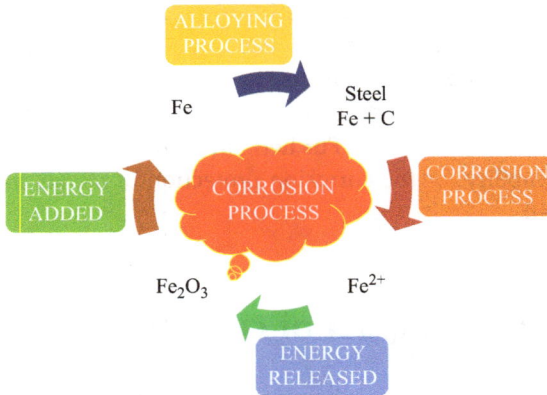

Figure 2.1: Schematic representation of a corrosion cycle.

- Expensive maintenance operations such as painting.
- Decreased effectiveness: Corrosion-induced encapsulation of heat transfer tubing and pipelines limits heat transmission and piping performance.
- Product wastage as a result of a rusted receptacle may result in serious mishaps and dangers.
- Practices to ensure security in the event of a fire, explosion, or dangerous product release.
- Pollution: Corrosion products can pollute chemicals, medications, colors, general merchandise, and other commodities, with disastrous repercussions for customers.
- Nuclear dangers: Health issues, such as lead pollution in drinking water, are possible as a result of deterioration. The Chernobyl accident is an ongoing instance of radioactive corrosion products being transported in water, killing humans, animals, and biological life.

The severity of corrosion would be determined by the susceptibility of certain metallic materials and alloys to a specific circumstance. Copper, for instance, oxidizes rapidly in the existence of ammonia, which is a severe concern in agricultural fields. Many antique bronze statues have been ruined by ammonia emitted by fertilizers. Atmospheric endurance is one way of corrosion protection and regulation, as is the application of inhibitors, as described in the following section.

2.3 Corrosion inhibitors

Corrosion is a deleterious and stealthy phenomenon. It causes issues for both large and small businesses. Since corrosion is unavoidable, it can be mitigated rather than prevented by using certain anticorrosion methods [16]. Inhibition is a widely researched topic in the realm of corrosion. Inhibition is a preventive strategy used to protect metallic components against corrosive environments. There are numerous efficient techniques for protecting metal from rust. However, one of the most well-known approaches to prevention against corrosion is the practice of CIs rather than the different measures to minimize or preclude metal surface damage or disintegration. Chemical substances, when introduced in modest proportions to a harsh environment, can reduce deterioration of the bare metal. Inhibitors are essential in a sealed ecological system with a lot of activity to maintain an appropriate and regulated dosage of the inhibitor [17]. CI dosages can range from 0.0001 to 1.5 wt% (1–15,000 ppm), and the efficiency of CI is calculated by the following formula:

$$E_f = \frac{R_i - R_o}{R_o} \times 100 \qquad (2.1)$$

where E_f is the efficiency of CI, R_i is the rate of corrosion in the presence of CI, and R_o is the rate of corrosion in the presence of inhibitor.

These characteristics can be achieved in a multitude of scenarios, comprising chilling water recirculation networks, oil generation, petroleum purification, and acid scouring of structural steel. Antifreeze for vehicle radiators is among the most well-known usage for inhibitors. Inhibitors can be organic or inorganic chemicals, and they are often dispersed in aqueous solutions. Because of the inexpensive cost and practice approach, employing inhibitors is the next step after standing up. CIs can take any shape (solids, liquids, and gases). CIs are chosen to inhibit corrosion based on their solubility or disposability in fluids and have been discovered to be an excellent and adaptable method of corrosion prevention as shown in Figure 2.2.

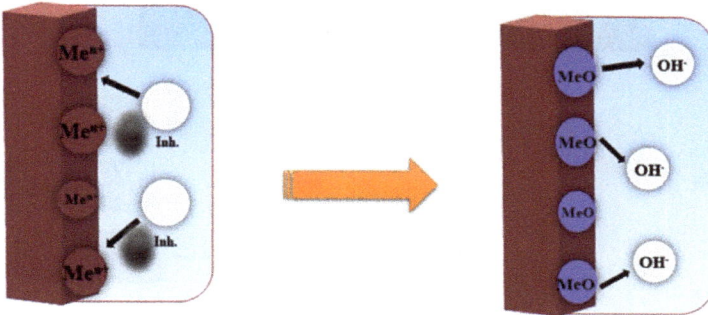

Figure 2.2: Process of corrosion inhibition (figure adapted with permission from ref. [18]; Copyright The Authors, some rights reserved; exclusive licensee [IntechOpen]. Distributed under a Creative Commons Attribution 3.0 License (CC BY) https://creativecommons.org/licenses/by/3.0/).

The usage of chemical inhibitors to slow down corrosion reactions varies greatly. Petroleum manufacturing and exploration, petroleum decontaminating, chemical production, H_2O purification, and item enhancement enterprises all utilize CIs. CIs have historically been regarded as the first line of defense in the extraction of oil, refining, and chemical sectors. A large number of laboratory investigations have been conducted on CIs by several researchers [19, 20].

2.4 Classification of corrosion inhibitors

Because corrosion is a surface reaction, adding a very little amount of CI to an interfacial region can avoid or minimize the corrosion process of a material subjected to a hazardous condition. As illustrated below, there are three primary forms of corrosion inhibition:

- Adsorption: The inhibitor is organically coated on the metallic surface, forming a robust, resilient thin layer with inhibitory properties.
- Surface layer: To safeguard the metallic platform, an oxide exterior film is formed on its outermost surface.
- Passivation: It occurs when the inhibitor combines with corrosive components in the aqueous phase, resulting in the formation of defensive reaction products.

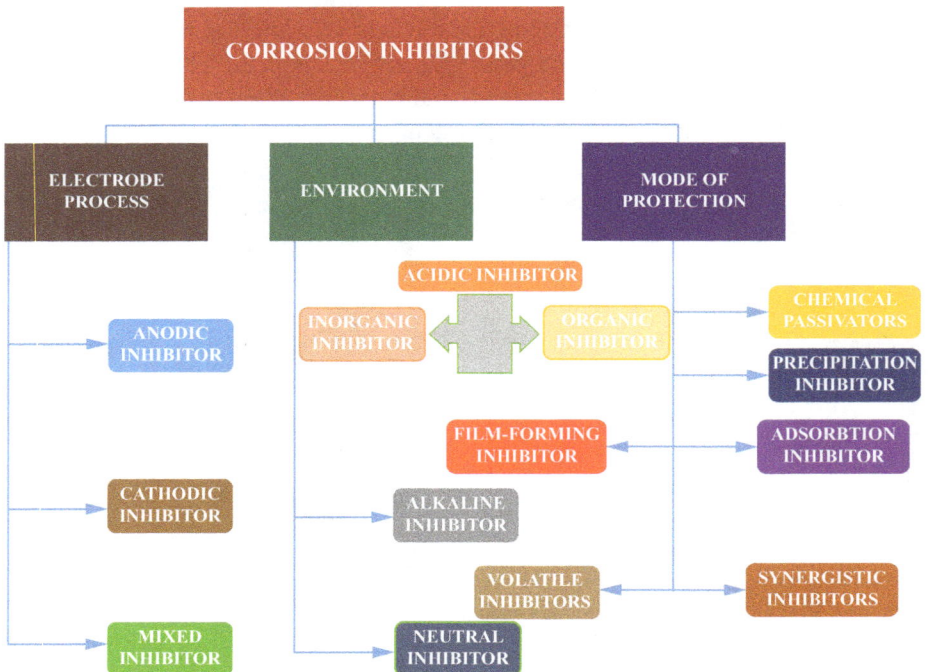

Figure 2.3: Classification of corrosion inhibitors.

CIs can be manmade or natural compounds, and based on the CIs' mechanism (described above), electrode process, type of corrosive media, and protection mechanism, they can be categorized as illustrated in Figure 2.3.

2.4.1 Based on electrode process

2.4.1.1 Anodic inhibitors

Anodic inhibitors work by obstructing the anode process and reinforcing the spontaneous instinct of passivation of the metallic outer layer, as well as by producing a coating adsorbed on the metallic substrate. They are also known as passivation inhibitors. In particular, CIs interact with the first generated corrosion product, culminating in a coherent and persistent layer on the metal substrate [21–23]. Anodic inhibitors, like chromates, phosphates, tungstates, and some other ions of transition metals with an extreme oxygen content, suppress the anodic corrosion process by creating a sparingly soluble product with afresh formed metal ion. A potentiostatic polarization curve of a mixture containing an inhibitor with anodic effectiveness is depicted in Figure 2.4.

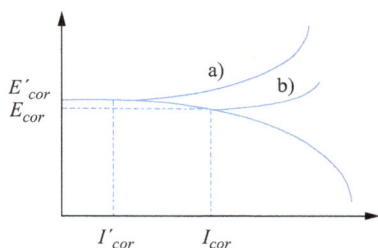

Figure 2.4: Potentiostatic polarization curves of a metallic substance in a solution in the (a) presence and (b) absence of anodic inhibitor (figure adapted with permission from ref. [12]; Copyright The Authors, some rights reserved; exclusive licensee [IntechOpen]. Distributed under a Creative Commons Attribution License https://creativecommons.org/licenses/by/3.0/).

CIs influence the anodic reaction, causing the metal's redox potential to switch to much more positive numbers. Anodic inhibitors interact with the metal cations (Men^+) formed on the anode terminal, generating hydroxides that are insoluble and coated on the metallic substrate, thus, impenetrable to ionic species. The decomposition of inhibitors produces OH^- ions. When the inhibitor levels are elevated sufficient, the cathodic current density at the main passivation voltage exceeds the threshold anodic current density, shifting the voltage for a sublime notion and, as a result, the material is usually coated. It is critical for the anodic inhibitors' efficacy that the inhibitor dosages in the environment be considered sufficient. An insufficient supply of inhibitors inhibits the creation of layer shielding because it does not encompass the entire metallic, allowing vulnerable areas of the metal and triggering localized corrosion [24, 25]. Concentration levels below the level of significance are far worse than none at all. In fact, they can cause pitting owing to anodic compression when compared to cathodic, or can aggravate deterioration, like widespread disintegration, due to sheer complacency breakdown. Anodic inhibitors include a variety of inorganic inhibitors such as nitrates, sodium chromates, phosphates, hydroxides, orthophosphates, and silicates. These agents are among the most efficacious and, as a result, are the most extensively used. Chromate-grounded compounds are the most affordable and potent inhibitors and have previously been used in a wide variety of applications. such as gasoline engines, rectifier diodes, and cooling devices. Anodic inhibitors are classified into two categories.

– Oxidizing anodic inhibitors: Those oxidizing anions are referred to as oxidizing passivates that can inactivate steel in the lack of O_2; for example, chromates, nitrites, and nitrates.
– Nonoxidizing anodic inhibitors: Nonoxidizing ions are those passivates that necessitate the oxygen content to passivate steel; for example, phosphates, tungstate, and molybdates.

Even though this form of management is effective, it can be risky because strong regional attacks can emerge if some regions are left uncovered due to inhibitor exhaustion. Even though anodic inhibitors are frequently utilized, a number of them have unfavorable properties. When such inhibitors are used at extremely trace quantities, they stimulate deterioration like pitting; hence, anodic CIs are labeled as harmful. As a result, it is critical to keep a close eye on the inhibitor concentration during the entire procedure.

2.4.1.2 Cathodic inhibitors

Cathodic CIs hinder the metal's cathodic processes from occurring during the electrochemical reaction. These inhibitors contain metal ions that can cause a cathodic reaction in the presence of alkalinity, resulting in insoluble complexes that deposit preferentially on cathodic sites. Install a dense and tenacious coating around the metallic outer layer, limiting the propagation of reducible species in these locations [26]. Hence, the surface impedance and transport limitation of the active groups, that is, O_2 diffusion and electrons conductive in these places, increase. Excessive cathodic inhibition is caused by these inhibitors. Figure 2.5 depicts an illustration of a metal polarization curve in solution with a cathodic CI. When the cathodic process is compromised, the corrosion potential shifts to greater negative numbers.

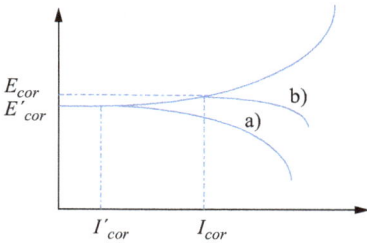

Figure 2.5: Potentiostatic polarization curves of a metallic substance in a solution in the (a) presence, and (b) absence of cathodic inhibitor (figure adapted with permission from ref. [12]. Copyright The Authors, some rights reserved; exclusive licensee [IntechOpen]. Distributed under a Creative Commons Attribution License https://creativecommons.org/licenses/by/3.0/).

The cathodic inhibitors envelop the metal by creating a barrier of insoluble precipitates. As a result, even if the metal is immersed, the metal's interaction with the environment is limited, limiting the corrosion reaction from occurring. Hence, cathodic inhibitors are dosage-independent, making them far more stable than anodic inhibitors. Cathodic inhibitors work in three ways to give inhibition: (a) cathodic poisons, (b) cathodic precipitates, and (c) oxygen scavengers. By incorporating the cathodic poisons, the cathodic reduction progression is managed by limiting H_2 reintegration and the form of protective effluent, but the metal's susceptibility to hydrogen-induced cracking increases. However, inorganic cathodic inhibitors include magnesium and nickel ions, which combine with the hydroxyl (OH^-) of moisture to generate insoluble ions in aqueous solution ($Mg(OH)_2$, $Ni(OH)_2$) that are precipitated on the cathodic region of the metal substrate, safeguarding it. Tannins, lignin [27], and calcium salts are further kinds of compounds that exhibit a similar reaction mechanism. It has also been observed in hard waters a type of inhibitory function attributed to the dominance of Mg or Ca bicarbonate on it. When transient hard water runs over the metallic substances, it can aid in the precipitation of carbonates, permitting processes to be close to equilibration and accumulates to occur on the metallic surface. These deposits, like $CaCO_3$, safeguard the metal by covering the cathodic region. As a result, these cathodic inhibitors are dependent solely on the composition of the water and not on the metallic constitution [21]. Antimony, arsenic, and bismuth oxides, and salts, for instance, are accumulated on the cathode area in an acidic medium. These cathodic inhibitors reduce the emission of hydrogen ions caused by voltage spikes, a phenomenon that can make hydrogen discharge problematic. Moreover, oxygen scavengers aid in corrosion prevention by inhibiting cathodic depolarization produced by oxygen. At room temperature, Na_2SO_3 (sodium sulfite) is most likely the most often employed oxygen sequester.

2.4.1.3 Mixed inhibitors

CIs are referred to as mixed ones since they hinder both the anodic and cathodic mechanisms involved in the corrosion process [28]. They are commonly film-forming chemicals that lead to precipitates to develop on the interface, indirectly inhibiting both anodic and cathodic points. Anodic inhibitors are often hazardous dampers, particularly when their dosages are too low. Cathodic inhibitors, on the other hand, are usually innocuous. Mixed inhibitors are less harmful than untainted anodic inhibitors, and they may not enhance corrosion severity in some instances. The silicates and phosphates are the most prevalent inhibitors in this class.

Gao et al. [29] used an IVIUM electrochemical workstation to examine the polarization curvatures of lignin-methacrylatoethyl trimethyl ammonium chloride (DMC) at various concentrations. Figure 2.6 depicts the polarization curves produced by the electrochemical model parameters. It demonstrated that the increase of the substance does not affect the dissolution or evolution of metal anode or cathode hydrogen, respectively. By producing a defensive layer on the outer layer of the carbon steel, the corrosion inhibition process only suppresses the active product of this reaction. With varying amounts of the suppressor, the Tafel slope of the negative electrode of the polarization curve enhanced from 94 to 154 mV, and the Tafel slope of the counter electrode improved from 114 to 185 mV, implying that the CI had a suppressive activity on anode and cathode interactions of Fe; however, the cathode Tafel slope risen significantly than the anode Tafel slope, indicating that lignin-DMC is a mixed-type inhibitor.

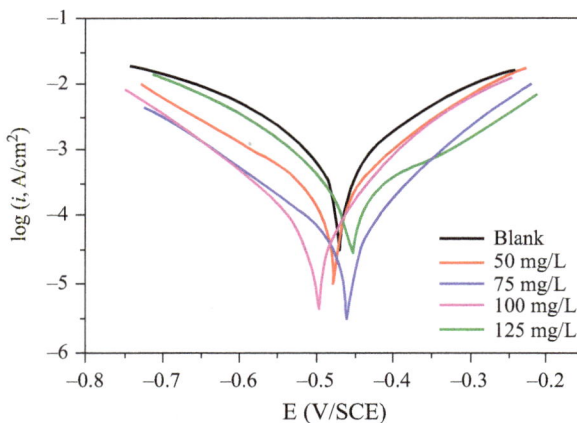

Figure 2.6: Polarization curves in acidic (1.0 mol/L HCl) solution containing various inhibitor concentrations at room temperature (figure adapted with permission from ref. [29]. Copyright The Authors, some rights reserved; exclusive licensee [MDPI]. Distributed under a Creative Commons Attribution (CC BY) license (http://creativecommons.org/licenses/by/4.0/)).

2.4.2 Based on environment

2.4.2.1 Acidic environment inhibitors

Acid CIs, particularly for Fe and steel, are widely used in a variety of sectors (e.g., petroleum field). Acidic CIs may be organic or inorganic. As acidic inhibitors, predominantly organic inhibitors are used because passivating inorganic inhibitors can be harmful in acidic conditions and instigate harsh regional attacks once the inactive layer is destroyed.

2.4.2.2 Inorganic corrosion inhibitors

In acidic media, chemicals like As_2O_3 and Sb_2O_3 have been observed to act as inhibitors. The safety in this instance is attributable to the diminution of electrophilic ions and accumulation on the metal substrate, as well as the decrease of the overvoltage of the primary cathodic depolarization process [30]. It has recently been discovered that the introduction of ions, namely, Pb^{2+} (lead), Ti^+ (titanium), Mn^{2+} (manganese), and Cd^{2+} (cadmium, heavy metal ions) inhibits iron degradation in acidic environment. Owing to the detrimental effect of passivating inorganic inhibitors in an acidic environment, they potentially cause regional assault once the passive layer is dissolved; hence, organic inhibitors are preferred particularly in acidic environments.

2.4.2.3 Organic corrosion inhibitors

Because of the utility of OCIs (organic CIs) in a wide range of sectors, they are an appealing study topic. The durability of the generated chelate determines the inhibitor's efficacy, and the inhibitor compound must have core sites capable of sustaining bonds with the outer surface of the metal via electrochemical reactions. Organic inhibitors can have both anodic and cathodic functions in the proximity of a disintegrating material, but when administered in adequate quantity, organic inhibitors influence the entire area of the oxidizing metal. Organic inhibitors, also known as film-forming inhibitors, shield the metallic substance by producing a water-repellent layer on its outer layer. Although film production is an adsorption procedure, system temperature and pressure are critical. Organic inhibitors will be absorbed based on their ionic charge and the charges on the substrate. Based on whether the metal is negatively or positively charged, cationic inhibitors including amines or anionic inhibitors including sulfonates will be deposited predominantly. For soluble organic inhibitors, the intensity of the adsorption link is the most important component. These compounds form a defensive coating of deposited entities on the metallic surface, acting as a stumbling block against metal putrefying in the solution. The following elements impact the effectiveness of an organic inhibitor:

- morphology, like the diameter of the organic compound'
- aromaticity and/or conjugated bonding, such as the chain length of carbon (C);
- bond type (either σ or π) and extent of bond formation in the chemical compound;
- essence and charges of the metal outer layer of the metal of adsorption configuration, such as adhesiveness to the metallic adsorbent;
- capability for a stack to develop dense or cross-linked [31].

The performance of these OCIs is connected to the existence of polar substituents containing S, O, or N heteroatoms, heterocyclic structures, and π-electrons, which usually include ionizable hydrophilic or hydrophobic sections [32]. The polar function is commonly thought of as the processing core for establishing the adherence behavior [33, 34]. Organic acid inhibitors including oxygen, nitrogen, and/or sulfur are deposited on the metal outer layer, obstructing vigorous corrosion centers. Thus, organic compounds are utilized as CIs in a multitude of industries, such as the oil and gas industry, water transfer, cooling process, furnaces, paint manufacturers, metal pipes, and construction materials. Benzalkonium chloride and rhodamine are two examples. Phosphate ions are commonly used in the infrastructure design [35]. The majority of organic inhibitors are deposited into the metal surfaces by shifting water molecules and generating a pressure shield. The presence of nonbonding electrons (one pair) and π-electrons in the inhibitor molecules expedite the allocation of electrons from the inhibitor to the metallic substance [36]. The efficacy of the inhibitor is determined by the stability of the chelate produced, which is determined mostly by the kind and character of the substitutes available in CI [37]. Although molecules with π-bonds are the most powerful and effective organic inhibitors, they have biological toxicities and are hazardous to the environment [38]. In this context, there has been an increase in the investigation of organic green CIs (GCIs), for instance, plant excerpts, gums, essential oils, and decontaminated ingredients to obtain environmentally beneficial substances. Organic GCIs are categorized as ionic liquids (ILs), surfactants, sustainable polymers, amino acids, medicinal chemicals, and substances taken from plants [39]. The existence of heteroatoms and an electron cloud in conjugation resulted in an outstanding inhibitory performance of organic GCIs, as evidenced by the publications. Bashir et al. [40] studied *Origanum vulgare* for its corrosion inhibition performance in an acidic solution for various metals. The results revealed a fortification percentage of 97.7% for an Al specimen and 91.4% for mild steel (MS) in 10^3 ppm. Adsorption fits in the Langmuir isotherm in both circumstances. Also, Bashir et al. [41, 42] studied the effects of phenylephrine on aluminum and MS. The rusting process was also elucidated, and it was shown that inhibitors slow both cathodic and anodic processes, thus acting as a mixed-type inhibitor. Moreover, alkyl dimethyl isopropyl ammonium hydroxide cationic surfactants for steel were produced, characterized, and examined by Badawia et al. [43]. The efficacy of inhibition was found to be greater than 96%.

Additionally, Messali [44] further confirmed the corrosion inhibition behavior of *N*-alkyl pyridinium ILs in acidic media, demonstrating an extreme of up to 95% at 0.01 M. Even though the metallic outer layer occupied is proportionate to the inhibitor concentrations, the inhibitor concentration levels in the medium are crucial [45]. "Amines, ascorbic acid, succinic acid, tryptamine, caffeine, and natural material excerpts are a few examples." However, some inhibitors continue to perform in the vapor phase (volatile CI, VCI). "Di-isopropyl ammonium nitrite or benzoate, ethanolamine benzoate or carbonate, and the blend of urea and $NaNO_2$" are a few examples.

2.4.2.4 Alkaline inhibitors

Metals that create amphoteric oxides are corrosive in basic environments. Many organic molecules are frequently utilized as metal inhibitors in alkaline solutions [46]. Altaf et al. studied the anticorrosion impacts of different metal cations (Al, Sn, Pb, and Mn) as dopants and five azoles (benzotriazole, mercaptobenzothiazole (MBT), benzimidazole, mercaptobenzimidazole, and thiadiazole) as an external layer on pure copper and four brass alloys in borate buffer at pH 10.4. Because of the availability of dopant ions, the corrosion rate (CR) of all brasses was determined to be leisurely than that of pristine Cu (0.132 mm/year), and MBT showed the highest corrosion prevention effectiveness between 91% and 96% for distinct brasses [47]. Because of the development of metal substances, chemicals for instance thiourea, substituted phenols, naphthol, and β-diketone have also been utilized as efficient inhibitors in basic solutions.

2.4.2.5 Neutral inhibitors

Inhibitors that work well in acid medium do not work well in neutral solutions because the mechanisms in the two components are distinct [48–50]. Zhuang et al. synthesized Gemini imidazoline surfactants from a variety of saturated fatty acids and investigated their corrosion inhibition behavior in NaCl solution for X70 carbon steel using electrochemical impedance spectroscopy, polarization curve, and quantum chemistry approach. The results show that such surfactants have an excellent inhibitory effect on carbon steel X70 in NaCl solution and that Gemini imidazoline has a higher restricted performance in an alkaline solution than in a neutral solution [51]. In neutral solutions, inhibitors interact with the oxide-covered surface of the metal, preventing oxygen reduction reactions at the cathode surface. These inhibitors safeguard the substrate surface from abrasion. In near-neutral solutions, certain surface-active chelating inhibitors have been considered to be effective inhibitors [52].

2.4.3 Based on mode of protection

2.4.3.1 Precipitation inhibitors

Precipitation-persuading inhibitor molecules are layer-materializing chemicals with a broad influence on the metal substrate, indirectly inhibiting both anodic and cathodic regions. Precipitation inhibitors are chemicals that enable precipitates to develop on the outer layer of a metallic substrate, forming a defensive coating. Hard H_2O with a significant calcium and magnesium content is far less abrasive than soft water because the salts in the hard water prefer to concentrate on the metal's surface and form a defensive barrier. The silicates and phosphates are the utmost prevalent competitors in this group. Many residential water softeners, for instance, employ sodium silicate to inhibit the development of rust water [53]. Sodium silicate safeguards steel, copper, and brass in aerated heating systems [54]. However, shielding may not always be accurate and is largely dependent on pH and a saturation index that varies according to H_2O makeup and temperature. Phosphates, moreover, need O_2 to be inhibited effectively. Silicates and phosphates do not provide the same level of fortification as chromates and nitrites, but they are extremely valuable in cases where harmless additions are necessary.

2.4.3.2 Adsorption inhibitors

This is the most commonly utilized category of inhibitors. In particular, they are organic molecules that become deposited on the surface of the metal and create a swamping effect throughout the exposed area, including cathodic and anodic situations. In general, they affect both cathodic and anodic processes similarly, but this is not always the situation. These are broadly employed in the acid preserving of hot rolled items to remove the black mill scale and are consequently referred to as curing inhibitors. Complexes with lone pairs of electrons, like nitrogen atoms in amines and quinolines, sulfur atoms in thio complexes, and oxygen atoms in aldehydes, are some significant examples.

2.4.3.3 Synergistic inhibitors

This is a single inhibitor used in cooling water systems. To obtain higher corrosion-resistant properties, the conjunction of inhibitors (anodic and cathodic) is frequently utilized. Chromate phosphates, polyphosphate silicates, zinc tannins, and zinc phosphates are a few examples. Furthermore, ILs have the critical characteristics of acting as a synergistic inhibitor. Because ILs and organic salts have this property, they could be employed as novel organic inhibitors, ideally with synergistic interactions. Several studies have been conducted to study the use of biologically acceptable anions and cations to create that could proceed toward the efficiency of chromates. Chong et al. [55] have described a class of ILs and organic salts that address dual functionality by containing both anions and cations and have demonstrated efficient suppression. The imidazolinium cation was combined with carboxylate anions to form these salts. The imidazolinium has a striking resemblance to imidazolium, except for the C_4–C_5 double bond saturation on the imidazolinium's central ring. These compounds were discovered to have intriguing physical features such as easy ionic conduction and synergistic corrosion suppression on MS depending on the type of anion in the salts. The effect of pH on the corrosion inhibiting the efficiency of the organic salt for MS in chloride settings was explored in this study. However, because of IL limitations, such synergy is not always possible. One of the main problems of ILs is that they have poor transport qualities, with higher viscosity and interfacial tension than standard organic solvents.

2.4.3.4 Film-forming inhibitors

Unlike adsorption inhibitors, which generate a straightforward deposited coating of the inhibiting component, several chemicals are known as film-forming inhibitors purport to prevent corrosion by generating an obstructing or shield film of a component other than the real inhibiting species itself. Such materials are usually unique to either the cathode or the anode. Cathodic film formation inhibitors are most commonly found in zinc and calcium salts. Benzoate is a prominent example of anodic film-forming inhibitors, which prevent corrosion throughout long-distance journeys. Xie et al. [56] investigated the corrosion behavior of aluminum metal matrix composites using uniformly distributed GNPs as protracted CIs. The findings demonstrate that acoustic composites without the synthesis of Al_4C_3 have a lower rate of corrosion attributed to the generation of a protective oxide film comprising GNP. Furthermore, uniformly distributed GNPs generated by extreme deformation of the material leads to improved abrasion resistance and a much more uniform protective coating covering [56].

2.4.3.5 Volatile corrosion inhibitors

VCIs are chemicals that are delivered to the location of decomposition via volatilization from a reservoir in a closed environment. They are also referred to as vapor-phase inhibitors (VPIs). Volatile basic chemicals like morpholine or hydrazine are delivered with mist in boilers to avoid weathering in condenser hoses by nullifying acidic CO_2 or moving interface pH away from excessive acidic and corrosive levels. Volatile solids including salts of dicyclohexylamine, cyclohexylamine, and hexamethylene amine are employed in closed vapor environments like shipping containers.

When these salts come into touch with a metal surface, their vapor shrinks and is hydrolyzed by any moisture to release protecting ions. It is preferable for a proficient VCI to stipulate inhibition quickly and for it to extant for an elongated time. Both properties are determined by the volatility of these chemicals, with immediate action necessitating high volatility and long-term protection necessitating low volatility.

2.4.3.6 Chemical passivators

Chemical passivators are compounds that have a comparatively elevated equilibrium potential (redox or electrode potential) and a reasonably low overpotential when they achieve passivity. Nitrites, for example, are utilized as CIs in antifreeze chilling waters. Chromates are primarily utilized as CIs in swirling chilled water systems. Zinc molybdate is often utilized in paints as a deterring ingredient.

2.4.3.7 Green corrosion inhibitors

There are no widely agreed definitions of "green" or "environmentally friendly" CIs. In practice, corrosion inhibition investigations have shifted to focus on personal health and the environment. Because of their low cost, renewable nature, and biodegradability, researchers have focused on the use of sustainable and environmentally friendly chemicals to achieve this goal, like plant excerpts and obsolete innocuous pharmaceuticals, which contain many organic components. Greener substitutes can be generated by restructuring current products or exploring novel chemistries for generating environmentally friendly products, as well as by minimizing discharge materials in terms of toxic effects and biodegradability. The corrosion and bioaccumulation costs are broken down. Toxicity can appear during the compound's manufacture or during its application, which must be considered while creating GCIs. Green substitutes for dangerous substances include amino acids, alkaloids, pigments, and tannins. Because of their biodegradability, eco-friendliness, competitive prices, and ease of availability, excerpts of some plants species and medicinal plants, as well as their derivatives, have been investigated as CIs for several metals and alloys under a wide range of environmental conditions. Karki et al. [57] examined the corrosion inhibition of *Equisetum hyemale* stem extract against MS in 1.0 M H_2SO_4 solution. The investigations demonstrated that it serves as an effective inhibitor and also verifies that the IE of EHE is greater than 85% at 1,000 ppm, which improves with dosage and declines with temperature rise [57]. However, Abdallah et al. evaluated the use of natural nutmeg oil as a GCI for carbon steel in a 1.0 M HCl solution. Chemical, electrochemical, and computational research revealed that nutmeg oil is an environmentally acceptable inhibitor of carbon steel-type L-52 (CS L-52) solution corrosion. These results also revealed that nutmeg oil was indeed a mixed inhibitor, with a percentage inhibitory efficacy of

94.73% at 500 ppm. Furthermore, the Langmuir isotherm was confirmed to be followed by the adsorption process. Moreover, Nathiya et al. demonstrated the enormous possibility of isolating *Dryopteris cochleata* extract of leaves for corrosion prevention of aluminum in 1 M H_2SO_4 [58]. When contrasted to aqueous extracts, methanol extract of *D. cochleata* leaves seems to be a more potent inhibitor. Alfred and colleagues, as well as James et al., investigated the inhibitory activities of *Swietenia macrophylla* leaf and Red Onion skin harvest on zinc corrosion in 0.5 M HCl solution. Because of the presence of the active metabolite quercetin, the inhibitory performance of red onion peel extract was examined to be 70%. Additionally, Akbar Ali Samsath Begum et al. [59] examined the CI activity of *Spilanthes acmella* aqueous leaves extract on steel specimen at various temperatures in a 1.0 M HCl solution [59]. The findings demonstrated that the occurrence of phytochemical components with functional groups such as electronegative heteroatoms like N, O, and S in the harvest adsorbed on the surface of the metal is fully accountable for the inhibitor's performance improvement, which encourages barrier properties by obstructing active surface area on the material as shown in Figure 2.7.

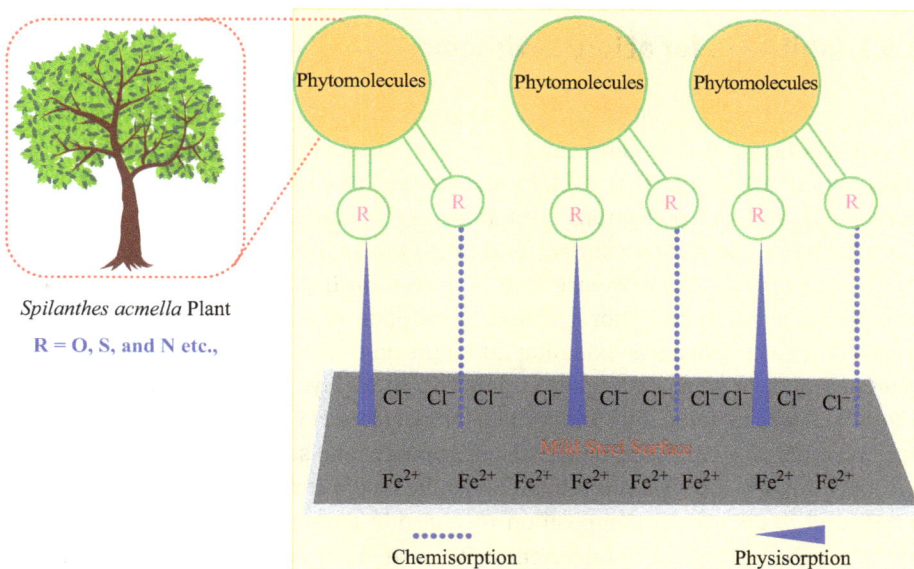

Figure 2.7: A schematic depiction of a possible corrosion inhibition process of *Spilanthes acmella* aqueous leaves extract (figure adapted with permission from ref. [59]. Copyright The Authors, some rights reserved; exclusive licensee [MDPI]. Distributed under a Creative Commons Attribution (CC BY) license (http://creativecommons.org/licenses/by/4.0/)).

2.5 Mechanism of corrosion inhibition

The bulk of inhibitor compound usage for aqueous/partially aqueous environments is associated with four basic environmental categories:
- Aqueous acid formulations are used in metal-cleaning operations like pickling to remove corrosion or mill scaling during metallurgical processes and manufacturing, or in post-service treatment of the metallic surface.
- Near-neutral pH (5–9), natural waters, supplying waters, and refrigeration and air-conditioning waters.
- Direct and indirect production of oil, as well as processing and transportation procedures.
- Atmospheric or gaseous corrosion in enclosed spaces, such as during transportation, warehousing, or any other constrained activity.

The following sections explain corrosion phenomena and the type of inhibitors used in terms of these four major conditions.

2.5.1 Inhibitors for acidic environments

Pickling, surface treatments, gas/oil well carburization, Zn, and Ni electroplating are all obvious benefits. It is generally understood in this discipline that inhibitors are often organic molecules that contain heteroatoms like sulfur (S), nitrogen (N), and oxygen (O), and also p electrons from aromatic structures, double and triple bonds. They adhere to the metallic surface and protect it from hostile attacks. Under some circumstances (e.g., the existence of H_2S or even during zinc electroplating), corrosion is characterized by major hydrogen absorption in acidic solutions. It has been shown that electrochemical oxidation inhibitors do not always restrict hydrogen consumption and, in some situations, enhance it. In reality, good S-containing CIs like thiourea, thioacid salts, and thiocyanates may create HS ions during acid degradation (possibly also catalyzed by the disintegrating metal substrate), hence increasing hydrogen ingress. Corrosion inhibitors in acidic media can engage with metallic materials and impact the decomposition response in several different behaviors, and few of which could operate concurrently. Because the method of activity of an inhibitor might fluctuate depending on the experimental parameters, it is often impossible to attribute a solitary association mode of action to it. As a result, the principal mechanism of an inhibitor may fluctuate based on variables like its dosage, the pH of the acidic moiety introduced, the type of the acid's anion, the existence of other entities in the system, the degree of reactivity to generate 2° (secondary) inhibitor molecules, and the metal's composition. The mode of accomplishment of CIs with the very same substituents may also differ depending on parameters like the influence of chemical structure on the charge distribution of the substituents and the length of the molecule's hydrocarbon component. In practice, inhibited hydrochloric acid has been shown time and again to become the most effective way of fouling removal. To understand the thermodynamics of fouling clearance, four equations are required. The following first three equations (2.2)–(2.4) reflect cathodic processes and one anodic process (eq. (2.5)):

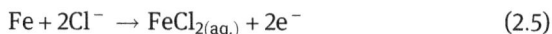

$$Fe_2O_3 + 4Cl^- + 6H^+ + 2e^- \rightarrow 2FeCl_{2(aq.)} + 3H_2O \qquad (2.2)$$

$$Fe_2O_4 + 6Cl^- + 8H^+ + 2e^- \rightarrow 3FeCl_{2(aq.)} + 4H_2O \qquad (2.3)$$

$$2H^+ + 2e^- \rightarrow H_2 \qquad (2.4)$$

$$Fe + 2Cl^- \rightarrow FeCl_{2(aq.)} + 2e^- \qquad (2.5)$$

According to these calculations, the basic iron acts as a reducer to speed up the dissolving of Fe oxides. Since determining the threshold for the dissolving of ensnaring oxides is challenging, an inhibitor is usually included for protection. Both anodic and cathodic inhibitors could be used to delay the deterioration of the metallic surfaces; subsequently, the snarling oxides have been dissolved.

However, considerable endeavors have been made to develop GCIs for acidic environments, which are primarily taken from plants, specifically seeds, leaves, and flowers. The inhibitory effects of these naturally present chemicals are linked to the existence of heterocyclic elements, such as alkaloids and cellulose, which ordinarily facilitate film development over the metal surface, hence, promoting protection against corrosion [60]. Furthermore, while plant derivatives are usually regarded benign for human beings and the environment, many very dangerous poisons are simply derived from plants, implying the need to analyze their toxicity toward the environments where they must be administered in any case. Because of resemblances in structural features with conventional modulators, data analysis investigations have demonstrated new categories of decreed nontoxic organic substances with considerable corrosion inhibiting characteristics. These chemicals like vitamins, amino acids, and medicines are also categorized by the occurrence of carbocyclic and heterocyclic structures, which are often flavorsome.

2.5.2 Corrosion inhibitors for oil/gas production and transport

Components of production of oil and gas and transportation infrastructures such as well tubing, drilling equipment, pipes, and sludge multiprocessors operate in extreme conditions of high temperature, pressures, and load in the presence of harmful substances such as CO_2, H_2S, and saline formation water. Corrosion is mostly sustained by proton depletion since the good condition is aerobic. However, when weak acids are present, weathering is assisted by the elimination of protonated anionic species, resulting in a higher oxidation potential than would be anticipated on a pH value basis. Furthermore, common procedures such as supplementary resource extraction by reinjection of carbon dioxide and water, and also various well-acidizing techniques to accelerate well expansion, exacerbate the harshness of operational apparatus-corrosive conditions [61]. For all of these considerations, the petroleum industry is the largest consumer of CIs. Because of these combinations, low-cost carbon and low-alloy steels can be used for tubing and pipelines, with corrosion-resistant compositions retained for the most severely stressed components. Efforts to eliminate dangerous chemicals like chromates and dichromates, as well as reduce the use of extremely hazardous amine chemicals have led to substantial advances in this field. When used in conjunction with saline-produced water such as CO_2 and H_2S, traditional regulators frequently contain active ingredients like amides, amines, imidazoline, and fatty acid analogs. Increased oil demand in recent decades has made it possible to reach deep-oil sources characterized by extraordinarily high temperatures and pressures. Since imidazoline rings dissolve at temperatures above 250 °C, corrosion regulators that are temperature resistant are necessary. Amido amines produced from fatty acids have increased thermostability and continue to be suppressive up to 320 °C. In the petroleum sector, ongoing research investigates the suppressive capacities of PLONOR chemicals in search of nontoxic inhibitors related to ecological authority criteria. Some of them have been demonstrated to be pretty much equivalent, if not better, to those of highly hazardous conventional formulations [62]. Appropriate alternative categories such as ILs and soluble polymeric molecules are being investigated as potential "green" and long-lasting CIs. ILs are cation–anion pairs (salts) that comprise both organic and inorganic constituents and have a melting point (MP) below 100 °C. They have new beneficial chemical and physical features, such as low vapor pressure and incredibly high resilience throughout a wide pH, temperature, and potential range. They are mixed-type CIs that work through chemisorption to safeguard the metallic coating. Ionic solutions based on imidazole have mostly been utilized to combat steel corrosion. Polymeric compounds with low ecological impact, for example, polyamino acids and carboxymethyl cellulose have exhibited intriguing inhibitory capabilities. About conventional small-molecule CIs, they have several adsorption sites per molecule and superior layer-forming properties, leading to an improvement in protective barrier qualities.

2.5.3 Inhibitors in near-neutral solution

Metal decomposition in neutral environments varies from that in acidic solutions in two significant ways. The predominant cathodic process in air-saturated solutions in neutral mixtures is the lessening of dissolved O_2, but in the acidic medium, it is H_2 generation. Disintegrating different metals in the acidic medium are oxide free; however, metal surfaces in neutral mixtures are blanketed with oxide, hydroxide, or salt coatings due to the lower solubility of these molecules. Due to these distinctions, chemicals that limit rusting in low pH solution by adsorption on oxide-free substrates often do not control deterioration in neutral solution. The anions of weak acids are common inhibitors for near-neutral solutions, with chromate, silicate, borate, and so on being perhaps one of the most significant in practice. Passivating oxide coatings on metals provide significant opposition to metal ionic movement and hinder the anodic process of metal dissolution. These inhibitive anions are known as anodic inhibitors and are often commonly utilized than cathodic inhibitors to prevent the dissolution of Fe, Zn, Al, Cu, as well as their alloys in near-neutral medium. The performance of inhibitive negatively charged ions on metal weathering in near-neutral medium includes the subsequent critical acts:

- The degradation speed of the passivating oxide layer is reduced.
- Regeneration of the oxide film through oxide reconstruction stimulation.
- Regeneration of the oxide film via pore trapping with insoluble complexes.
- Adsorption of hostile anions is avoided.

The most important of these responsibilities seems to be the lowering of the decomposition pace of the passivating oxide covering. Negatively charged inhibitive ions more likely generate a surface compound with the metallic ion of the oxide (i.e., iron(III), zinc(II), and aluminum(III)), which is more stable than comparable compounds with water, hydroxyl ions, or combative anions. Blocking in neutral solutions can also be caused by the deposition of chemicals that can create or stabilize a protective layer on a metallic substrate. By precipitation or reactivity, the CI may produce a surface coating of an insoluble salt. These salt layers, which are frequently pretty dense and may even be apparent, obstruct passage to the metal substrate, especially of dissolved oxygen. They are often referred to as cathodic inhibitors because they have poor electric conductivity, and O_2 decline does not ensue on the layer's interface.

2.5.4 Gaseous and atmospheric corrosion

VCIs are a low-cost and effective means of avoiding ambient or air corrosion of metallic materials and alloys. The stimulation of the atmosphere with the minute dosage of inhibitive substance to generate a beneficial impact is the basis for volatile corrosion inhibition. A VCI molecule must be volatile as well as encourage electrochemical phenomena such as changing the voltage in the diffuse region of the double layer, which regulates the movement of electrode reaction components. The potential of a VPI to access the outer layer to be shielded is the first need for good accuracy. The second need is that the molecular transmission rate not be too sluggish to preclude an early exposure on the MS by the corrosive media prior the inhibitor can operate. These two requirements are connected to the vapor pressure of the inhibitor molecule, the range between the inhibitor source(s) and the metallic materials, and the availability of the substrates [63]. Volatile complexes fast achieve protective vapor concentrations, but in the event of non-airtight cages, inhibitor utilization is considerable and the adequate protection time is limited. Low vapor pressure inhibitors do not degrade quickly and can provide longer-lasting prevention. However, achieving a shielding vapor intensity requires longer effort. Additionally, deterioration may happen during the preliminary saturation phase, and if the area is not hermetically coated an appropriate inhibitor dosage may by no means be attained. As a result, the chemical component utilized as a volatile inhibitor must have a vapor pressure that is neither too high nor too low, but rather some ideal vapor pressure [64]. Notably, the highly efficient VCIs are the results of a weak volatile alkali reacting with a weak volatile acidic solution. While ionized in aqueous environments, such compounds endure significant breakdown, the degree of which is relatively liberated of dosage. In the instance of "amine nitrites and amine carboxylates," the remaining consequence of those processes is as follows.

$$H_2O + R_2NH_2NO_2 \rightarrow (R_2NH_2)^+ : OH^- + H^+ : (NO_2^-) \tag{2.6}$$

The type of the deposited film generated at the steel–water contact is a critical component in determining VCI efficiency. Different metals sensitized to VCI vapors in sealed bags show signs of being blanketed by a water-repellent-adsorbed film. The contact angle of pure water on these substrates grows with the contact period. Experiments on the deposition of VCIs from the gaseous phase support the hypothesis that VCIs interact with the metallic outer layer, giving protection against decomposition.

2.6 Application of corrosion inhibitors

Though the harmful repercussions of corrosion can be greatly reduced by selecting extremely anticorrosive elements, the cost element connected with the similar favors the usage of low-priced metallic substances in conjunction with excellent corrosion management technologies for several commercial purposes. Corrosion inhibitors are important in this regard, either as standalone inhibitors or as a constituent in chemical formulations. They have a broad range of industrial uses, including chilling waters, petroleum sectors, and electronics. A vast range of CIs has been proposed and implemented for deployment to diverse arrangements based on the material administered, the kind of interface that is corrosive, the form of deterioration experienced, and the circumstances to which the material is subjected [65]. The efficacy and utility of a CI in one number of scenarios do not always suggest a similar alternative. Several applications, like industrial water purification plants, use a blend of many CIs as well as other chemicals, namely, antiscalants, biocides, and polymeric dispersants are delivered. CIs often function in tandem with improved coatings and cathodic protection. The following are some of the most common applications for CIs.

2.6.1 In cooling waters

Chilling waters are employed in a variety of service sectors, including petroleum refineries, steel plants, petrochemical industries, electricity generation plants, food processing units, and chemical processing plants. However, degradation in open cooling water equipment is intimately connected to the development of mineral scales, solid accumulation, and microbial ensnarling. As a result, a complete cooling water supplement formulation contains antiscaling compounds, polymeric diluents, and biocides, as well as CIs to safeguard iron (Fe), copper (Cu), aluminum (Al), and their alloys. Corrosion in open systems is exacerbated by high levels of salt and dissolved oxygen. Another important issue is galvanic corrosion. Successful water treatment is particularly cost-effective since it reduces inspection shutdowns and allows for the extensive use of low-carbon steel. Following the elimination of poisonous chromates and nitrates, these antiscaling and anticorrosive compositions now include more environmentally benign silicates, molybdates, and polyphosphate/phosphonates. To prevent calcium phosphate deposition, entangling, and eutrophication of oceans, ponds, and lakeside, their phosphate/polyphosphate concentration is kept low. Phosphates, as opposed to polyphosphates, are more resilient to break down throughout a larger pH and temperature range. Furthermore, when combined with Zn or Fe ions, they safeguard metal surfaces more competently through chemical adsorption, eventually form firm metal substances, and adsorb on particulate matters, resulting in enhanced antiscaling and dispersion capabilities. Low-phosphorus all-organic synergistic compounds, such as phosphono-tricarboxylates and fatty amines and phosphonocarboxylates and polyvinyl pyrrolidone, have lately demonstrated exceptional characteristics while having a lesser environmental impact [66, 67]. Their effectiveness is primarily due to the creation of a defensive layer composed of adsorbed moieties of organic compounds combined with iron oxides or hydroxides.

2.6.2 In concrete

Reinforcing steel bars in well-constructed and handled appropriately masonry stay inactive attributed to the prevalence of a basic environment within the holes with a pH of almost 13 or more. Corrosion, on the other hand, might begin as a result of chloride infiltration or concrete carbonation. Chloride ions destroy the inactive coating locally, resulting in the localized assault, whereas carbon dioxide interacts with the moist cement matrix, reducing the pH, indicating steel inactivity degradation, and generating a highly universal corrosion shape. CIs can be an effective and cost-effective technique to prevent rebar disintegration in harsh environments. Nitrite, in the form of calcium salts, is a common corrosion deterrent for concrete. Calcium nitrate is remarkably efficient versus Cl^- attack, while sodium (Na) and lithium (Li) salts can be employed in restoration mixtures and are significantly efficient at inhibiting carbonation consequences. They have oxidant properties and hinder corrosion initiation more effectively than transmission. Many other compounds have recently been explored as CIs. In basic medium imitating chloride-contaminated concrete pore electrolyte, sodium b-glycerophosphate (GPH) inhibited steel corrosion well, and in GPH-impregnated mortar, it provided steel rebar with transient impedance to corrosion over chloride penetration [68]. Moreover, "salts of monocarboxylic and polycarboxylic acids" were evaluated in chloride-polluted alkaline conditions for their inhibitory capabilities against embrittlement of steel reinforcing bars. The findings confirmed that electron-loving moieties, such as amine or hydroxyl constituents, boosted the inhibitory activity of these compounds, with lactate being the highly effective inhibitor [69].

2.6.3 Acid pickling and boilers

To protect the metallic substances from being attacked by the acidic media in which the metal is scrubbed of mill scale (bark lamination), and to also inhibit the consequent evolution of H_2, inhibitors are introduced, which need to be dissolved or diffused in the fluid. Examples include thiourea, amino, and its derivatives, and propargyl alcohol. Nevertheless, thermoelectric boilers, in particular, use ammonia, cyclohexylamine, alkanol, and morphline as inhibitors in several operations. To avoid pipe corrosion, the inhibitors are also introduced by HCl used for limescale solubilization.

2.6.4 Oil industry

To minimize the detrimental effect of CO_2, H_2S, and organic acids, sodium carbonates or organic amine complexes are used, allowing the usage of cheap and low corrosion-resistant components in crude oil extraction wells. Sulfonated oils and sodium nitrite are used in the pipelines for gasoline and kerosene. "Fatty amines, fatty acids, imidazolines, and quaternary ammonium salts are all depleted by oil well." Internal pipe pitting corrosion in moist gas transmission owing to condensation of dispersed corrosive gases in water. Corrosion is caused by the dissolving of abrasive gases like carbon dioxide and hydrogen sulfides, and also by the humidification of acid vapors.

2.6.5 Electronic industry

Chemical mechanical polishing (CMP) is a method that removes particles from the interface of a microelectronic gadget wafer and planarizes the area. CMP, in its most basic form, is introducing slurry, such as an aggressive and active chemical solution, to a buffing pad that buffs the area of a microelectronic gadget wafer to accomplish the ablation, planarization, and polishing operations. There are numerous papers published on CMP solutions that use CIs. Yoshikawa et al. invented a mechanochemical buffing mixture consisting of "phosphonium salt, triazole-based CI, and colloidal silica [70]." Also, sarcosine and its salts were licensed by Chang et al. as CIs while cleaning following chemical mechanical sharpening [71]. Corrosion inhibiting coatings comprising "benzotriazole or nitrophenylhydrazine (for metals such as W-Cu, Mo-Cu, and Cu)" in the wiring coatings, coated at least on the periphery of the plating stacks are used in licensed printed circuit boards with long-term dependability [72].

2.7 Patents in the field of corrosion inhibitors

- Subramaniyan et al. obtained a patent for a polyisobutylene phosphorus sulfur compound that acts as a CI for elevated temperature naphthenic acid electrochemical reaction, as well as sulfur corrosion as a result, and the performance improves while phosphorus requirements are reduced [73].
- Khomyakova et al. patented a chemical composition containing "*n*-brominebenzal -*m*-nitroaniline, 2-chlorine-6-diethylamino-4-methyl pyridine, 1,3-bis(carbamoyl-tio)-2-(*N*, *N*-die methylamine) propane hydrochloride, and urotropin" for use in metallic cleaning and acid remediation of energy, culinary, and medicinal facilities. The formulation is helpful for steel, aluminum, and nickel preservation, as well as reducing steel hydrogenation [74].
- Seetharaman et al. investigated the corrosion inhibiting characteristics of tetrazole-based inhibitors. The invention, was patented for CI compounds and their compositions. These substances premised on tetrazole were effective in combating or suppressing the deterioration of surfaces used in cooling water systems [75].
- Ekoue-Kovi et al. studied the CI properties of 2-mercaptobenzimidazole derivatives. The innovation, which was patented for preventing metal corrosion, is generally related to procedures and compounds. More particularly, the process entails contacting a hydrocarbon-containing liquid with an effective amount of a composition to avoid metal corrosion [76].
- Rane et al. researched the corrosion inhibition of a metal surface in contact with an aqueous system and were patented for employing 2-substituted imidazoles and 2-substituted benzimidazoles, and provide increased corrosion protection of metals in the aqueous system. The approach employs CIs that are generally resistant to halogen attack and give good corrosion resistance in the presence of oxidizing halogen-based biocides [77].
- Mandal et al. were granted a patent for their innovation, which relates to a low-cost, high-performance CI compound and the method of making it. An ester of cashew nut shell liquid, a dimer fatty acid, a long-chain amine, and a solvent make up the CI. The CI formulation protects the inside metal surface of pipes against corrosion caused by petroleum products [78].

– Hulbert et al. were granted a license to publish the forest fire-retardant composition, which includes at least one anhydrous salt and at least one hydrate salt. The final diluted product suppressed, retarded, and controlled forest fires while demonstrating corrosion resistance and low toxicity [79].

2.8 Conclusion

Corrosion is one of the most serious and widespread challenges experienced in the commercial world. Various fluids that are contained in machinery are used in manufacturing sectors, but when these equipment corrodes, it poses a serious threat to life and property. The concern is inescapable, and it will have a comparable impact as natural catastrophes. As a result, the industrial business sector is required to invest tens of billions of dollars each year. Nonetheless, while resolving this issue is impossible, taking preventative measures to keep the metal surface from corroding is more cost-effective. The application of CIs is among the most efficient strategies for shielding metal surfaces against corrosion, owing to its low cost and ease of use. In this chapter, a complete examination of corrosion and the usage of several CIs is discussed. CIs are a highly efficient way of corrosion prevention. Awareness of the process mechanism simplifies the selection of CIs, increases performance, prevents procedure impairment, and reduces side effects. In this regard, the classification of CIs depending on the electrode process, corrosive media, and so on has also been documented. Moreover, the chapter provides a brief idea about the corrosion mechanisms, how it occurs in different conditions, its impact, and how to calculate the inhibitory efficiency to determine CR. Additionally, recent patents in the sector of corrosion inhibition are evaluated and documented in this chapter. Since many organic, inorganic, and polymeric substances performed well, however, many of them are hazardous and do not meet all environmental regulations. So, regarding toxicity, biodegradability, and biomagnification, inhibitors must be examined in the context of wellness, security, and environmental consequences. These factors contribute to the creation of innovative inhibitors with little or no environmental impact, which are referred to as eco-friendly or GCIs. However, their quantities should also be evaluated on a regular schedule, and deficits should be replenished either by adequate inhibitor supplements or by total fluid resuscitation, as advised, for instance, with engine coolants. Wherever possible, constant monitoring should be employed, but consider the fact that information from surveillance equipment, sensors, coupons, and so on about the behavior of that particular segment at that particular system component.

References

[1]   Bentiss F, Lagrenee M, Traisnel M, Hornez JC. Corrosion inhibition of mild steel in 1 M
        hydrochloric acid by 2, 5-bis (2-aminophenyl)-1, 3, 4-oxadiazole. Corrosion, 1999; 55:
        968–976.
[2]   Yurt AY, Balaban A, Kandemir SU, Bereket G, Erk B. Investigation on some Schiff bases as HCl
        corrosion inhibitors for carbon steel. Materials Chemistry and Physics, 2004; 85: 420–426.
[3]   Sharma S, Ganjoo R, Kr. Saha S, Kang N, Thakur A, Assad H, Sharma V, Kumar
        A. Experimental and theoretical analysis of baclofen as a potential corrosion inhibitor for
        mild steel surface in HCl medium. Journal of Adhesion Science and Technology, 2021; 16:
        1–26.
[4]   Sharma S, Chaudhary RS. Inhibitive action of methyl red towards corrosion of mild steel in
        acids. Bulletin of Electrochemistry, 2000; 16: 267–271.
[5]   Assad H, Kumar A. Understanding functional group effect on corrosion inhibition efficiency of
        selected organic compounds. Journal of Molecular Liquids, 2021; 344: 117755.
[6]   Hanoon M, Zinad DS, Resen AM, Al-Amiery AA. Gravimetrical and surface morphology studies
        of corrosion inhibition effects of a 4-aminoantipyrine derivative on mild steel in a corrosive
        solution. International Journal of Corrosion and Scale Inhibition, 2020; 9: 953–966.
[7]   Al-Amiery AA, Kadhum AA, Mohamad AB, Musa AY, Li CJ. Electrochemical study on newly
        synthesized chlorocurcumin as an inhibitor for mild steel corrosion in hydrochloric acid.
        Materials, 2013; 6: 5466–5477.
[8]   Al-Amiery AA, Kadhum AA, Mohamad AB, Junaedi S. A novel hydrazinecarbothioamide as a
        potential corrosion inhibitor for mild steel in HCl. Materials, 2013; 6: 1420–1431.
[9]   Thakur A, Kumar A. Sustainable inhibitors for corrosion mitigation in aggressive corrosive
        media: A comprehensive study. Journal of Bio- and Tribo-Corrosion, 2021; 7: 1–48.
[10]   Sanyal B. Organic compounds as corrosion inhibitors in different environments – A review.
        Progress in Organic Coatings, 1981; 9: 165–236.
[11]   Pourbaix M. Applications of electrochemistry in corrosion science and in practice. Corrosion
        Science, 1974; 14: 25–82.
[12]   Dariva CG, Galio AF. Corrosion inhibitors – Principles, mechanisms and applications.
        Developments in Corrosion Protection, 2014; 16: 365–378.
[13]   El Bribri A, Tabyaoui M, Tabyaoui B, El Attari H, Bentiss F. The use of Euphorbia falcata extract
        as eco-friendly corrosion inhibitor of carbon steel in hydrochloric acid solution. Materials
        Chemistry and Physics, 2013; 141: 240–247.
[14]   Cisse MB, Zerga B, El Kalai F, Touhami ME, Sfaira M, Taleb M, Hammouti B, Benchat N, El
        Kadiri S, Benjelloun AT. Two dipodal pyridin-pyrazol derivatives as efficient inhibitors of mild
        steel corrosion in HCl solution – Part I: Electrochemical study. Surface Review and Letters,
        2011; 18: 303–313.
[15]   Geethamani P, Kasthuri PK, Aejitha S. Mitigation of mild steel corrosion in 1 M sulphuric acid
        medium by Croton sparciflorus – A green inhibitor. Chemical Science Review and Letters,
        2014; 2: 507–516.
[16]   Bashir S, Singh G, Kumar A. An investigation on mitigation of corrosion of aluminium by
        Origanum vulgare in acidic medium. Protection of Metals and Physical Chemistry of Surfaces,
        2018; 54: 148–152.
[17]   Kumar A, Thakur A. Encapsulated nanoparticles in organic polymers for corrosion inhibition.
        Corrosion Protection at the Nanoscale, 2020; 345–362.
[18]   Palanisamy G. Corrosion Inhibitors. IntechOpen: London, 2019, 24.

[19] Bashir S, Lgaz H, Chung IM, Kumar A. Potential of Venlafaxine in the inhibition of mild steel corrosion in HCl: Insights from experimental and computational studies. Chemical Papers, 2019; 73: 2255–2264.

[20] Aejitha S, Kasthuri PK, Geethamani P. Comparative study of corrosion inhibition of Commiphora caudata and Digera muricata for mild steel in 1 M HCl solution. Asian Journal of Chemistry, 2016; 28: 307.

[21] Bashir S, Lgaz H, Chung IM, Kumar A. Effective green corrosion inhibition of aluminium using analgin in acidic medium: An experimental and theoretical study. Chemical Engineering Communications, 2020; 21: 1–0.

[22] Sharma S, Ganjoo R, Saha SK, Kang N, Thakur A, Assad H, Kumar A. Investigation of inhibitive performance of Betahistine dihydrochloride on mild steel in 1M HCl solution. Journal of Molecular Liquids, 2021; 347: 118383.

[23] Ganjoo R, Kumar A. Current trends in anti-corrosion studies of surfactants on metals and alloys. Journal of Bio-and Tribo-Corrosion, 2022; 8: 1–35.

[24] Muster TH, Hughes AE, Furman SA, Harvey T, Sherman N, Hardin S, Corrigan P, Lau D, Scholes FH, White PA, Glenn M. A rapid screening multi-electrode method for the evaluation of corrosion inhibitors. Electrochimica Acta, 2009; 54: 3402–3411.

[25] Bashir S, Singh G, Kumar A. An investigation on mitigation of corrosion of aluminium by Origanum vulgare in acidic medium. Protection of Metals and Physical Chemistry of Surfaces, 2018 Jan 1; 54(1): 148–152.

[26] Forsyth M, Seter M, Tan MY, Hinton B. Recent developments in corrosion inhibitors based on rare earth metal compounds. Corrosion Engineering, Science and Technology, 2014; 49: 130–135.

[27] El-Deeb MM, Ads EN, Humaidi JR. Evaluation of the modified extracted lignin from wheat straw as corrosion inhibitors for aluminum in alkaline solution. International Journal of Electrochemical Science, 2018; 13: 4123–4138.

[28] Raheem D. Effect of mixed corrosion inhibitors in cooling water system. Al-Khwarizmi Engineering Journal, 2011; 7: 76–87.

[29] Gao C, Wang S, Dong X, Liu K, Zhao X, Kong F. Construction of a novel lignin-based quaternary ammonium material with excellent corrosion resistant behavior and its application for corrosion protection. Materials, 2019; 12: 1776.

[30] Saei E, Ramezanzadeh B, Amini R, Kalajahi MS. Effects of combined organic and inorganic corrosion inhibitors on the nanostructure cerium based conversion coating performance on AZ31 magnesium alloy: Morphological and corrosion studies. Corrosion Science, 2017; 127: 186–200.

[31] Sanyal B. Organic compounds as corrosion inhibitors in different environments – A review. Progress in Organic Coatings, 1981; 9: 165–236.

[32] Brycki BE, Kowalczyk IH, Szulc A, Kaczerewska O, Pakiet M. Organic corrosion inhibitors. Corrosion Inhibitors, Principles and Recent Applications, 2018; 3: 33.

[33] Yaro AS, Khadom AA, Wael RK. Apricot juice as green corrosion inhibitor of mild steel in phosphoric acid. Alexandria Engineering Journal, 2013; 52: 129–135.

[34] Thakur A, Kumar A. A review on thiazole derivatives as corrosion inhibitors for metals and their alloys. European Journal of Molecular & Clinical Medicine, 2020; 7: 3702–3712.

[35] Yohai L, Vázquez M, Valcarce MB. Phosphate ions as corrosion inhibitors for reinforcement steel in chloride-rich environments. Electrochimica Acta, 2013; 102: 88–96.

[36] Goyal M, Kumar S, Bahadur I, Verma C, Ebenso EE. Organic corrosion inhibitors for industrial cleaning of ferrous and non-ferrous metals in acidic solutions: A review. Journal of Molecular Liquids, 2018; 256: 565–573.

[37] Kadhim A, Jawad RS, Numan NH, Al-Azawi RJ. Determination the wear rate by using XRF technique for Kovar alloy under lubricated condition. Power, 2017; 30: 17.

[38] El-Haddad MN. Chitosan as a green inhibitor for copper corrosion in acidic medium. International Journal of Biological Macromolecules, 2013; 55: 142–149.

[39] Sharma S, Kumar A. Recent advances in metallic corrosion inhibition: A review. Journal of Molecular Liquids, 2020; 322: 114862.

[40] Bashir S, Singh G, Kumar A. Shatavari (Asparagus Racemosus) as green corrosion inhibitor of aluminium in acidic medium. Journal of Materials and Environmental Science, 2017; 8: 4284–4291.

[41] Bashir S, Sharma V, Singh G, Lgaz H, Salghi R, Singh A, Kumar A. Electrochemical behavior and computational analysis of phenylephrine for corrosion inhibition of aluminum in acidic medium. Metallurgical and Materials Transactions A, 2019; 50(1): 468–479.

[42] Bashir S, Thakur A, Lgaz H, Chung IM, Kumar A. Computational and experimental studies on Phenylephrine as anti-corrosion substance of mild steel in acidic medium. Journal of Molecular Liquids, 2019; 293: 111539.

[43] Badawi AM, Hegazy MA, El-Sawy AA, Ahmed HM, Kamel WM. Novel quaternary ammonium hydroxide cationic surfactants as corrosion inhibitors for carbon steel and as biocides for sulfate reducing bacteria (SRB). Materials Chemistry and Physics, 2010; 124: 458–465.

[44] Messali M. A green microwave-assisted synthesis, characterization and comparative study of new pyridazinium-based ionic liquids derivatives towards corrosion of mild steel in acidic environment. Journal of Materials and Environmental Science, 2011; 2: 174–185.

[45] Sherif ES. Effects of 2-amino-5-(ethylthio)-1, 3, 4-thiadiazole on copper corrosion as a corrosion inhibitor in 3% NaCl solutions. Applied Surface Science, 2006; 252: 8615–8623.

[46] Noor EA. Potential of aqueous extract of Hibiscus sabdariffa leaves for inhibiting the corrosion of aluminum in alkaline solutions. Journal of Applied Electrochemistry, 2009; 39: 1465–1475.

[47] Altaf F, Qureshi R, Yaqub A, Ahmed S. Electrochemistry of corrosion mitigation of brasses by azoles in basic medium. Chemical Papers, 2019; 73: 1221–1235.

[48] Mistry BM, Patel NS, Sahoo S, Jauhari S. Experimental and quantum chemical studies on corrosion inhibition performance of quinoline derivatives for MS in 1N HCl. Bulletin of Materials Science, 2012; 35: 459–469.

[49] Khaled KF. Molecular simulation, quantum chemical calculations and electrochemical studies for inhibition of mild steel by triazoles. Electrochimica Acta, 2008; 53: 3484–3492.

[50] Umoren SA, Li Y, Wang FH. Synergistic effect of iodide ion and polyacrylic acid on corrosion inhibition of iron in H2SO4 investigated by electrochemical techniques. Corrosion Science, 2010; 52: 2422–2429.

[51] Zhuang W, Wang X, Zhu W, Zhang Y, Sun D, Zhang R, Wu C. Imidazoline Gemini surfactants as corrosion inhibitors for carbon steel X70 in NaCl solution. ACS Omega, 2021; 6: 5653–5660.

[52] Aejitha S, Asthuri PK. Geethamani, "Inhibition effect of Antigonon leptopus extract on mild steel in sulphuric acid medium,". Indian Journal of Applied Research, 2014; 4: 51–53.

[53] Kumar SH, Karthikeyan S. Amoxicillin as an efficient green corrosion inhibitor for mild steel in 1M sulphuric acid. Journal of Materials and Environmental Science, 2013; 4: 675–984.

[54] Wang C, Chen J, Hu B, Liu Z, Wang C, Han J, Su M, Li Y, Li C. Modified chitosan-oligosaccharide and sodium silicate as efficient sustainable inhibitor for carbon steel against chloride-induced corrosion. Journal of Cleaner Production, 2019; 238: 117823.

[55] Chong AL, Mardel JI, MacFarlane DR, Forsyth M, Somers AE. Synergistic corrosion inhibition of mild steel in aqueous chloride solutions by an imidazolinium carboxylate salt. ACS Sustainable Chemistry & Engineering, 2016; 4: 1746–1755.

[56] Xie Y, Meng X, Mao D, Qin Z, Wan L, Huang Y. Homogeneously dispersed graphene nanoplatelets as long-term corrosion inhibitors for aluminum matrix composites. ACS Applied Materials & Interfaces, 2021; 13: 32161–32174.

[57] Karki N, Neupane S, Chaudhary Y, Gupta DK, Yadav AP. Equisetum hyemale: A new candidate for green corrosion inhibitor family. International Journal of Corrosion and Scale Inhibition, 2021; 10: 206–227.

[58] Nathiya RS, Raj V. Evaluation of Dryopteris cochleata leaf extracts as green inhibitor for corrosion of aluminium in 1 M H2SO4. Egyptian Journal of Petroleum, 2017; 26: 313–323.

[59] Begum AA, Vahith RM, Kotra V, Shaik MR, Abdelgawad A, Awwad EM, Khan M. Spilanthes acmella leaves extract for corrosion inhibition in acid medium. Coatings, 2021; 11: 106.

[60] Umoren SA, Eduok UM. Application of carbohydrate polymers as corrosion inhibitors for metal substrates in different media: A review. Carbohydrate Polymers, 2016; 140: 314–341.

[61] Haruna K, Saleh TA, Quraishi MA. Expired metformin drug as green corrosion inhibitor for simulated oil/gas well acidizing environment. Journal of Molecular Liquids, 2020; 315: 113716.

[62] Sharma S, Kumar A. Recent advances in metallic corrosion inhibition: A review. Journal of Molecular Liquids, 2021; 322: 114862.

[63] Fiaud C. Theory and practice of vapour phase inhibitors. The Institute of Materials, Corrosion Inhibitors (UK), 1994; 1994: 1–1.

[64] Kumar A, Bashir S. Ethambutol: A new and effective corrosion inhibitor of mild-steel in acidic medium. Russian Journal of Applied Chemistry, 2016; 89: 1158–1163.

[65] Kumar A, Bashir S. Review on corrosion inhibition of steel in acidic media. International Journal of ChemTech Research, 2015; 8: 391–396.

[66] Ochoa N, Moran F, Pébère N, Tribollet B. Influence of flow on the corrosion inhibition of carbon steel by fatty amines in association with phosphonocarboxylic acid salts. Corrosion Science, 2005; 47: 593–604.

[67] Ochoa N, Moran F, Pébère N. The synergistic effect between phosphonocarboxylic acid salts and fatty amines for the corrosion protection of a carbon steel. Journal of Applied Electrochemistry, 2004; 34: 487–493.

[68] Monticelli C, Frignani A, Trabanelli G. Corrosion inhibition of steel in chloride-containing alkaline solutions. Journal of Applied Electrochemistry, 2002; 32: 527–535.

[69] Pastore T, Cabrini M, Coppola L, Lorenzi S, Marcassoli P, Buoso A. Evaluation of the corrosion inhibition of salts of organic acids in alkaline solutions and chloride contaminated concrete. Materials and Corrosion, 2011; 62: 187–195.

[70] Yoshikawa S, Uemura T, Ohashi H, Inaba T Composition for metal polishing. JP Patent. 2009, 117517.

[71] Chang SY, inventor; Uwiz Technology Co Ltd, assignee. Sarcosine compound used as corrosion inhibitor. United States Patent 2012, US 8,337,716.

[72] Sato H Printed circuit boards with excellent solder wettabilty and long term reliability in electrical and mechanical connection to electronic components. JP Patent. 2008, 109076.

[73] Subramaniyam M, inventor; Dorf Ketal Chemicals (I) Private Ltd, assignee. High temperature naphthenic acid corrosion inhibition using organophosphorous sulphur compounds and combinations thereof. United States Patent 2015, US 9,090,837.

[74] Khomyakova TA, Vostrikova DA, Kravtsov EE, Gerlov VS, Starkova NN, Ogorodnikova NP, Kondratenko TS Metal corrosion inhibitor in sulfuric and hydrochloric acids. RU Patent. 2009, 2343226.

[75] Seetharaman J, Reny EA, Johnson DA, Sawant KB, Sivaswamy V. *U.S. Patent No. 9,771,336.* U.S. Patent and Trademark Office: Washington, DC, 2017.

[76] Ekoue-Kovi K, Tomar N, Jadhav D, Peyton KB, Sorrells JL, Jones IM. *U.S. Patent No. 10,457,817.* U.S. Patent and Trademark Office: Washington, DC, 2019.
[77] Rane D, Seetharaman J, Atkins JM, Harbindu A, Anant P, Sivaswamy V, Cheruku P. *U.S. Patent No. 10,202,694.* U.S. Patent and Trademark Office: Washington, DC, 2019.
[78] Mandal T, Sharma M, Shanti P, Yadav A, Arora AK, Puri SK, . . . Suresh R. *U.S. Patent No. 10,563,114.* U.S. Patent and Trademark Office: Washington, DC, 2020.
[79] Hulbert D, Burnham RJ, Schnarr MS, Geissler G, Wilkening DW, McLellan J. *U.S. Patent No. 10,960,250.* U.S. Patent and Trademark Office: Washington, DC, 2021.

Omar Dagdag, Rajesh Haldhar, Seong-Cheol Kim,
Elyor Berdimurodov, Eno E. Ebenso, Savaş Kaya

3 Natural corrosion inhibitors: adsorption mechanism

Abstract: The motivation behind this chapter is to give an outline of the bibliographic insights expected to structure our segment. We examine modern green corrosion inhibitors and their restraint and adsorption systems. The corrosion repressing impacts of different plant extracts are additionally discussed utilizing weight loss (WL) and electrochemical techniques. Also, this chapter talks about surface investigation, including chemical examination.

Keywords: Natural products, corrosion inhibitor, adsorption mechanism, Langmuir, physisorption

3.1 Introduction

To secure or protect the metal, eco-friendly corrosion inhibitors are adsorbed on the surface of the metal, utilizing pai-electronic frameworks, sulfur, nitrogen, oxygen, and phosphorus [1]. This adsorption can occur in two unique manners; physisorption or chemisorption. Physisorption or physical adsorption is reversible and has a low adsorption enthalpy of around 20–40 kJ/mol [2–6]. Chemisorption or chemical adsorption is irreversible, with an enthalpy of about 80–240 kJ/mol [7]. Physisorption takes place at low temperatures and with increase in temperature, it decreases. Additionally, it has less activation energy. Chemisorption takes place at high temperatures and with increase in temperature, it increases [4, 8–11]. Chemisorption has comparatively higher activation energy. Corrosion is a natural process, where a metal starts rusting when it comes into direct contact with moisture. It combines hydrogen evolution (cathodic corrosion) and metal dissolution (anodic corrosion) [12, 13]. As per the reports by NACE (National Association of Corrosion Engineers), USD 2.5 trillion is lost each year (3.4% of world GDP) due to corrosion. Analyzing by countries, South Korea's losses are USD 1,198 billion, India's losses are USD 1,670 billion, China's losses are USD 9,330 billion, European Union's losses are USD 16,950 billion, Germany's losses are USD 3,593 billion, and Russia's losses are USD 2,113 billion, while Saudi Arabia's losses are USD 718 billion due to the corrosion of steel; all losses are per year [14]. Thus, corrosion is a huge economical issue. Therefore, it is necessary to prevent of steel from the corrosion process. Figure 3.1 shows the cost of corrosion in different sectors.

https://doi.org/10.1515/9783110760583-003

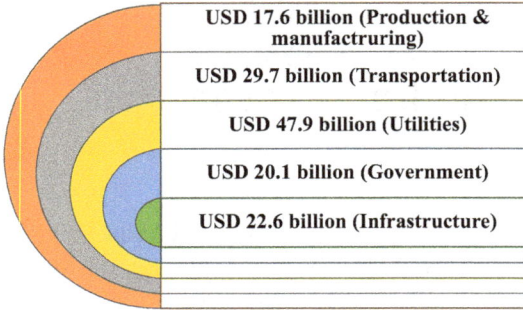

| USD 17.6 billion (Production & manufactruring) |
| USD 29.7 billion (Transportation) |
| USD 47.9 billion (Utilities) |
| USD 20.1 billion (Government) |
| USD 22.6 billion (Infrastructure) |

Figure 3.1: Cost of corrosion in different sectors.

Several methods are commercially available to resist steel corrosion, but most of those are non-eco-friendly and require a high budget as well. Some crop materials such as organic products, seeds, dry leaves, bark, and peel of some fruits have a non-toxic, non-hazardous, and eco-friendly nature [15–18]. These materials can be used as corrosion resistance specialists. Their easy availability and economic accessibility make them more favorable. Figure 3.2 shows the types of CIs.

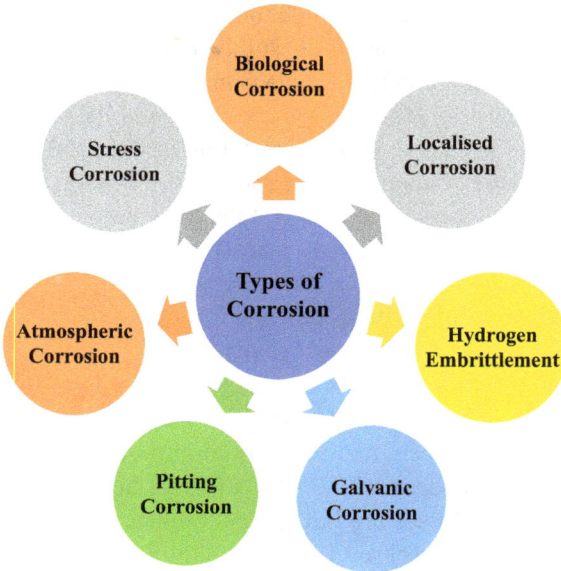

Figure 3.2: Different types of corrosion.

A study by the international organization, NACE, (2013) confirmed that the estimated cost of global corrosion was $ 2.5 trillion (3.4% of world GDP). As mentioned, this is largely the percentage of corrosion from economic services sectors of all countries. Corrosion costs were 20% for the US, 26% for India, 26% for Japan, 51% for Kuwait, and 20% for the United Kingdom in manufacturing (Figure 3.3). The global market for corrosion inhibitors was $6 billion in 2013, $7.7 billion in 2020, and it is estimated to reach about $10 billion in 2027 [19]. During the COVID-19 pandemic, consumer behavior has changed across all walks of life. On the other hand, industries will have to restructure their strategies to adapt to the demands of a changing market. Corrosion prevention is critical because it can destroy the economy and cause a slew of safety issues. Corrosion causes a variety of economic losses due to metal damage. From a security standpoint, this can be extremely dangerous. It also reduces the strength of parts in vehicles, aircraft, and ships, among other things. Corrosion of pressure vessels, boilers, and chemical containers can be extremely dangerous in the chemical industry. Economic corrosion causes two types of losses: direct and indirect [2]. Organic corrosion inhibitors, with unique electron pairs, include nitrogen, oxygen, sulfur, and phosphorus, and may also contain pi-electronic structural portions that interact with the metal and aid in the adsorption process. Several methods are commercially available to resist steel corrosion but most of those are non-eco-friendly and require a high budget as well. Thus, it is a huge economical issue as well. Therefore, it is necessary that

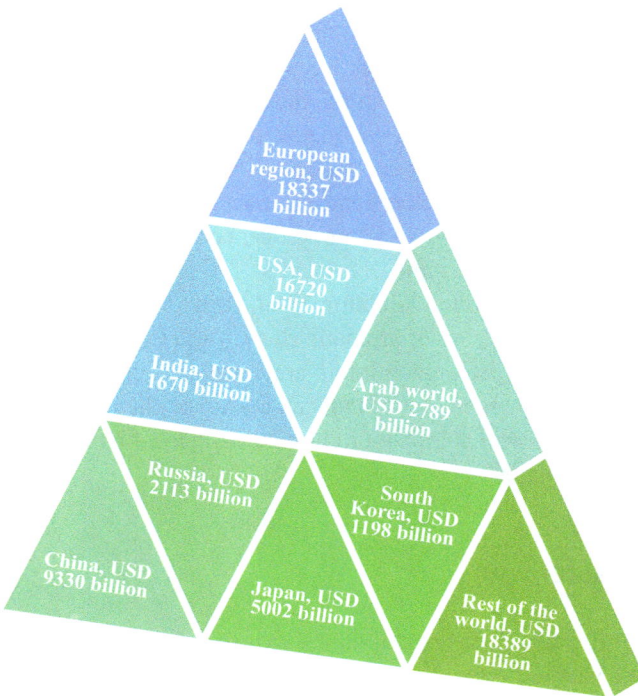

Figure 3.3: Cost of corrosion by major countries every year.

steel is protected from the corrosion process. Several methods are commercially available to resist steel corrosion but most of those are non-eco-friendly and require a high budget as well. Some crop materials, such as organic products, seeds, dry leaves, bark, and peel of some fruits have a non-toxic, non-hazardous, and eco-friendly nature. These materials can be used as corrosion resistance specialists. Their easy availability and economic accessibility make them more favorable [4]. Stringent ecological guidelines are significant factors in keeping down the development of the anticorrosion industry. Increasing worries about the effect and wellbeing of synthetic compounds have triggered severe limitations for the makers of consumption inhibitors. Manufacturers of erosion coatings currently urge them to choose nonpoisonous choices [20]. The use of non-harmful consumption inhibitors in harsh conditions makes it hard for organizations to change the plan norms. For instance, an added substance in the oil and gas industry is zinc phosphate. Albeit powerful, this inhibitor doesn't stick to the properties of chromate buildings, for example, hexavalent chromium. As per the United States Department of Labor, drugs cause leukemia in people and creatures.

3.2 The inhibition mechanism of the plant extract

The application is being examined to affirm the viability of anticorrosion activity of the plant-based extract as an economical and effective natural corrosion inhibitor [21]. As a rule, the utilization of a green inhibitor on the metal surface can be named physisorption, chemisorption, or a blend of both, called mixed adsorbents [22]. Mixed types of inhibitors offer better corrosion inhibition against both cathodic and anodic responses [23]. These kinds of corrosion inhibitors can be impacted by the idea of the nature of the consumption inhibitor, its substance properties, consumption opposition, and corrosion obstruction [24]. For example, Hanini et al. [25] fostered an actual model, dependent on the idea of consumption inhibitor. Their study revealed that the plant extract contains an enormous number of aromatic rings, various connections, and heteroatoms that may handily be protonated into a corrosive environment. This protonation brings about all-around charged groups that can adsorb on the metal alloy surface, which is positively charged with sulphate or chloride particles, framing contact spans called electrostatic interaction.

3.2.1 Weight loss measurement

The principle technique that can be utilized to decide the anticorrosion properties of the inhibitor is to look at the viability of the inhibition resistance at various temperatures, depending on weight-loss estimations. For instance [26], it has been seen that the corrosion resistance rate reduces with increasing temperature at any concentration of CCDE [26]. This condition is regularly deciphered as being symbolic of physical adsorption [27]. The adsorption models can predict utilizing the inclusion proportion, which is dependent on the utilization of various inhibitor concentrations and the degree of surface coverage, which is acquired from weight-loss estimations [26]. Table 3.1 shows the order of the quantity of general isothermal adsorption tests by the utilization of surfaces and other weight-loss esteems.

Table 3.1: Typical instances of adsorption isotherms.

Isotherm	Linear form
Langmuir (c)	$\dfrac{C_{inh}}{\theta} = \dfrac{1}{K_{ads}} + C_{inh}$
Temkin (a)	$\exp(-2f\theta) = K_{ads}*C_{inh}$
Freundlich (b)	$\log(\theta) = \log(K_{ads}) + n\log(C_{inh})$
Frumkin	$\dfrac{\theta}{1-\theta}\exp(-2f\theta) = b*C_{inh}$

For example, Verma et al. [28] developed information and distinctive isothermal models, as shown in Figure 3.4.

By utilizing the most reasonable isothermal model, as shown in Table 3.1, the K_{ads} esteem is then used to gauge the ΔG_{ads}^0, which is dependent on the situation condition:

$$\Delta G_{ads}^0 = -RT \ln(C_{inh}*K_{ads})$$

The physisorption or the physical adsorption is reversible and has a low adsorption enthalpy of around 20–40 kJ/mol. Chemisorption or chemical adsorption is irreversible, with an enthalpy of about 80–240 kJ/mol [29]. A negative sign indicates that the adsorption cycle has occurred spontaneously [30].

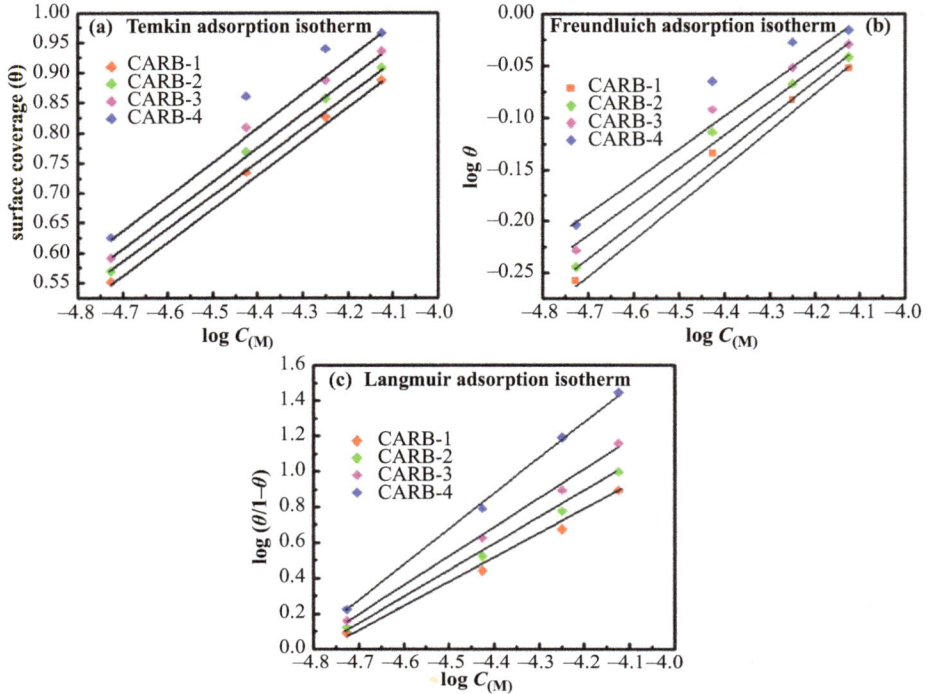

Figure 3.4: Diagram of the adsorption isotherm with various models. Adapted with permission from reference [28]. ACS Publications (2016) Distributed under a Creative Commons Attribution License 4.0 (CC BY).

3.2.2 Potentiodynamic polarization

Besides weight loss examination, the kind of inhibitor and the inhibition mechanism of corrosion inhibitors, in view of regular plant extract concentrates, are not entirely settled by polarization bends and inhibitor concentration, between the inhibitors, at room temperature. On the off chance that the inhibitor causes a huge change on the cathode breadth and on the anodic size of the polarization bend, the values of corrosion potential of E_{corr} is essentially steady; therefore, the inhibitor can be arranged based on the type of cathodic blend [24].

3.3 Chemical analysis

Despite theoretical estimation strategies, chemical investigations, for example, infrared and UV-vis. spectrometry, were performed to analyze the chemical properties of the inhibitor [31]. UV-vis. spectrometry was performed because an adjustment of the position of the absorbance maxima and an adjustment of the absorbance value demonstrate that a complex is probably going to form between two particles in the test media [32]. It tends to be utilized as an indicator of the development of a new product when the substrate is inundated in a destructive media containing an inhibitor. The energy and impact of the new band are not entirely set in stone; they can be varied by contrasting the individual spectra of the destructive medium, and the substrate and the arrangement of the substrate and the inhibitor. Another resulting examination is acquiring information through the Fourier-transform infrared (FT-IR) range about the different functional groups and vibrational modes of different molecules present in the plant extract. Mobin et al. [33] examined the avoidance of multifunctional bunches, zirconium acetate derivation and *Cissus quadrangularis* (CQ), in the consumption obstruction of steel and acidic conditions by the FT-IR test. They compared the spectra of the particular inhibitors that were adsorbed on the metal after a WL analysis, in which the adsorbed compounds were eliminated and permitted FT-IR examination. They, likewise, noticed that significant peaks, for example, OH for alcohol, OH for carboxylic acid, CO for carbonyl, CN for amine, and CO for ester were present toward the lower frequency. These perceptions affirm the association of CQ's phytochemicals to steel [33].

3.4 Surface analysis

The best information for deciding the surface morphology of substrates, with and without inhibitors, can be analyzed by the XPS trailed, followed by AFM and SEM. As a rule, an unprotected consumption inhibitor has a destructive surface because of the impact of the metal and the destructive surfaces. In any case, the region within the sight of the inhibitor is by and large smoother than the region without the inhibitor. It is by and large acknowledged that the arrangement of an adsorbent film on an inhibitor makes the metal smooth.

The consequence of SEM micrographs affirms the inhibitory impact of numerous natural inhibitors on the outer layer of carbon steel (CS) alloy. A corroded surface is framed when the substrate is covered with an acidic climate, without inhibitors. The EDS spectral with SEM micrographs additionally provide surface properties and compound wellbeing on a chosen surface, harmed by consumption. For instance [24], it is observed that without henna extract, the metal substrate was harmed by a corrosive medium. EDS has been utilized to show that a consumed surface is covered with more chloride than a surface extricated with henna [24].

Likewise, X-ray spectroscopy (XPS) is a surface assessment that has the exceptional capacity to affirm the presence of metal consumption inhibitors. XPS provides the recognizable proof of content with various oxidations and various information, utilizing the nonlinear least-squares curve-fitting technique [34]. For example, Liao et al. [35] studied the impact of ginger extract on the corrosion security of CS. They might assume a significant part in working on the activity of ginger extract on metals in the creation of carbonaceous organic films, as shown by XPS spectroscopy. The carbon content of the C–C/CH bond was, likewise, 77.03% of the solids content and the C–O obligation of the particles was 22.96%.

3.5 Conclusions

This chapter examined the current turn of events and the utilization of the different green corrosion inhibitors. The financial commitments of some key nations were discussed. The various kinds of corrosion processes and their inhibitor–metal mechanisms were additionally discussed. The utilization of various kinds of plant extracts as economical, effective, and efficient corrosion inhibitors has been broadly talked about.

Author's contributions: Omar DAGDAG wrote the manuscript; Rajesh Haldhar, Seong-Cheol Kim, and Elyor Berdimurodovc edited the manuscript; Eno E. Ebenso and Savaş Kaya done formal analysis, conceptualization, control, validation, and visualization.

References

[1]   Haldhar R, Prasad D, Bhardwaj N. Extraction and experimental studies of Citrus aurantifolia as an economical and green corrosion inhibitor for mild steel in acidic media. Journal of Adhesion Science and Technology, 2019; 33: 1169–1183.
[2]   Haldhar R, Prasad D, Mandal N, Benhiba F, Bahadur I, Dagdag O. Anticorrosive properties of a green and sustainable inhibitor from leaves extract of Cannabis sativa plant: Experimental and theoretical approach. Colloids and Surfaces. A, Physicochemical and Engineering Aspects, 2021; 614: 126211.
[3]   Haldhar R, Prasad D, Bhardwaj N. Experimental and theoretical evaluation of acacia catechu extract as a natural, economical and effective corrosion inhibitor for mild steel in an acidic environment. Journal of Bio-and Tribo-Corrosion, 2020; 6: 1–11.
[4]   Haldhar R, Prasad D. Corrosion resistance and surface protective performance of waste material of Eucalyptus globulus for low carbon steel. Journal of Bio-and Tribo-Corrosion, 2020; 6: 1–13.

[5]   Haldhar R, Prasad D, Kamboj D, Kaya S, Dagdag O, Guo L. Corrosion inhibition, surface
        adsorption and computational studies of Momordica charantia extract: A sustainable and
        green approach. SN Applied Sciences, 2021; 3: 1–13.
[6]   Haldhar R, Prasad D, Saharan H. Performance of Pfaffia paniculata extract towards corrosion
        mitigation of low-carbon steel in an acidic environment. International Journal of Industrial
        Chemistry, 2020; 11: 1–12.
[7]   Haldhar R, Prasad D, Saxena A, Singh P. Valeriana wallichii root extract as a green &
        sustainable corrosion inhibitor for mild steel in acidic environments: Experimental and
        theoretical study. Materials Chemistry Frontiers, 2018; 2: 1225–1237.
[8]   Saxena A, Prasad D, Haldhar R, Singh G, Kumar A. Use of Saraca ashoka extract as green
        corrosion inhibitor for mild steel in 0.5 M H_2SO_4. Journal of Molecular Liquids, 2018; 258:
        89–97.
[9]   Saxena A, Prasad D, Haldhar R. Investigation of corrosion inhibition effect and adsorption
        activities of Cuscuta reflexa extract for mild steel in 0.5 M H_2SO_4. Bio electrochemistry, 2018;
        124: 156–164.
[10]   Nam JG, Lee ES, Jung WC, Park YJ, Sohn BH, Park SC, et al. Photovoltaic enhancement of dye-
        sensitized solar cell prepared from [TiO_2/ethyl cellulose/terpineol] paste employing
        TRITON™ X-based surfactant with carboxylic acid group in the oxyethylene chain end.
        Materials Chemistry and Physics, 2009; 116: 46–51.
[11]   Haldhar R, Kim SC, Prasad D, Bedair M, Bahadur I, Kaya S, et al. Papaver somniferum as an
        efficient corrosion inhibitor for iron alloy in acidic condition: DFT, MC simulation, LCMS and
        Electrochemical studies. Journal of Molecular Structure, 2021; 1242: 130822.
[12]   Haldhar R, Prasad D, Bahadur I, Dagdag O, Kaya S, Verma DK, et al.. Investigation of plant
        waste as a renewable biomass source to develop efficient, economical and eco-friendly
        corrosion inhibitor. Journal of Molecular Liquids, 2021; 335: 116184.
[13]   Haldhar R, Prasad D, Saxena A. Myristica fragrans extract as an eco-friendly corrosion
        inhibitor for mild steel in 0.5 M H_2SO_4 solution. Journal of Environmental Chemical
        Engineering, 2018; 6: 2290–2301.
[14]   Haldhar R, Prasad D, Saxena A, Kumar R. Experimental and theoretical studies of Ficus
        religiosa as green corrosion inhibitor for mild steel in 0.5 M H_2SO_4 solution. Sustainable
        Chemistry and Pharmacy, 2018; 9: 95–105.
[15]   Haldhar R, Prasad D, Bhardwaj N. Surface adsorption and corrosion resistance performance
        of Acacia concinna pod extract: An efficient inhibitor for mild steel in acidic environment.
        Arabian Journal for Science and Engineering, 2020; 45: 131–141.
[16]   Haldhar R, Prasad D, Saxena A. Armoracia rusticana as sustainable and eco-friendly
        corrosion inhibitor for mild steel in 0.5 M sulphuric acid: Experimental and theoretical
        investigations. Journal of Environmental Chemical Engineering, 2018; 6: 230–5238.
[17]   Haldhar R, Prasad D, Bahadur I, Dagdag O, Berisha A. Evaluation of Gloriosa superba seeds
        extract as corrosion inhibition for low carbon steel in sulfuric acidic medium: A combined
        experimental and computational studies. Journal of Molecular Liquids, 2021; 323: 114958.
[18]   Haldhar R, Prasad D, Saxena A, Kaur A. Corrosion resistance of mild steel in 0.5 M H_2SO_4
        solution by plant extract of Alkana tinctoria: Experimental and theoretical studies.
        The European Physical Journal Plus, 2018; 133: 1–18.
[19]   Koch G, Varney J, Thompson N, Moghissi O, Gould M, Payer J. International measures of
        prevention, application, and economics of corrosion technologies study. NACE International,
        2016; 216: 2–3.
[20]   Saxena A, Prasad D, Haldhar R, Singh G, Kumar A. Use of Sida cordifolia extract as green
        corrosion inhibitor for mild steel in 0.5 M H_2SO_4. Journal of Environmental Chemical
        Engineering, 2018; 6: 694–700.

[21] Steinfeld B, Scott J, Vilander G, Marx L, Quirk M, Lindberg J, et al.. The role of lean process improvement in implementation of evidence-based practices in behavioral health care. The Journal of Behavioral Health Services & Research, 2015; 42: 504–518.

[22] Faisal M, Saeed A, Shahzad D, Abbas N, Larik FA, Channar PA, et al.. General properties and comparison of the corrosion inhibition efficiencies of the triazole derivatives for mild steel. Corrosion Reviews, 2018; 36: 507–545.

[23] Bryckilwon B, Kowalczyk I, Szulc A, Kaczerewska O, Pakiet M. Organic corrosion inhibitors. *Corrosion Inhibitors, Principles and Recent Applications*. IntechOpen, 2019; 3–33.

[24] Ostovari A, Hoseinieh S, Peikari M, Shadizadeh S, Hashemi S. Corrosion inhibition of mild steel in 1 M HCl solution by henna extract: A comparative study of the inhibition by henna and its constituents (Lawsone, Gallic acid, α-d-Glucose and Tannic acid). Corrosion Science, 2009; 51: 1935–1949.

[25] Hanini K, Merzoug B, Boudiba S, Selatnia I, Laouer H, Akkal S. Influence of different polyphenol extracts of Taxus baccata on the corrosion process and their effect as additives in electrodeposition. Sustainable Chemistry and Pharmacy, 2019; 14: 100189.

[26] Odewunmi NA, Umoren SA, Gasem ZM, Ganiyu SA, Muhammad Q. L-citrulline: An active corrosion inhibitor component of watermelon rind extract for mild steel in HCl medium. Journal of the Taiwan Institute of Chemical Engineers, 2015; 51: 177–185.

[27] Popova A, Sokolova E, Raicheva S, Christov M. AC and DC study of the temperature effect on mild steel corrosion in acid media in the presence of benzimidazole derivatives. Corrosion Science, 2003; 45: 33–58.

[28] Verma C, Olasunkanmi LO, Ebenso EE, Quraishi MA, Obot IB. Adsorption behavior of glucosamine-based, pyrimidine-fused heterocycles as green corrosion inhibitors for mild steel: Experimental and theoretical studies. The Journal of Physical Chemistry C, 2016; 120: 11598–11611.

[29] Rubaye A, Abdulwahid A, Al-Baghdadi SB, Al-Amiery A, Kadhum AAH, Mohamad A. Cheery sticks plant extract as a green corrosion inhibitor complemented with LC-EIS/MS spectroscopy. International Journal of Electrochemical Science, 2015; 10: 8200–8209.

[30] Özcan M, Solmaz R, Kardaş G, Dehri İ. Adsorption properties of barbiturates as green corrosion inhibitors on mild steel in phosphoric acid. Colloids and Surfaces. A, Physicochemical and Engineering Aspects, 2008; 325: 57–63.

[31] Alvarez PE, Fiori-Bimbi MV, Neske A, Brandan SA, Gervasi CA. Rollinia occidentalis extract as green corrosion inhibitor for carbon steel in HCl solution. Journal of Industrial and Engineering Chemistry, 2018; 58: 92–99.

[32] Abboud Y, Abourriche A, Saffaj T, Berrada M, Charrouf M, Bennamara A, et al.. 2, 3-Quinoxalinedione as a novel corrosion inhibitor for mild steel in 1 M HCl. Materials Chemistry and Physics, 2007; 105: 1–5.

[33] Mobin M, Basik M, Shoeb M. A novel organic-inorganic hybrid complex based on Cissus quadrangularis plant extract and zirconium acetate as a green inhibitor for mild steel in 1 M HCl solution. Applied Surface Science, 2019; 469: 387–403.

[34] Casaletto MP, Figà V, Privitera A, Bruno M, Napolitano A, Piacente S. Inhibition of Cor-Ten steel corrosion by "green" extracts of Brassica campestris. Corrosion Science, 2018; 136: 91–105.

[35] Liao LL, Mo S, Luo HQ, Li NB. Corrosion protection for mild steel by extract from the waste of lychee fruit in HCl solution: Experimental and theoretical studies. Journal of Colloid and Interface Science, 2018; 520: 41–49.

Ambrish Singh, Kashif Rahmani Ansari, Shivani Singh,
Mumtaz Ahmed Quraishi

4 Plants as corrosion inhibitors for metals in corrosive media

Abstract: The protection of metals from corrosion is widely carried out by adding green inhibitors, that is, natural plant extracts. The extracts were tested to contain corrosion, utilizing modes such as gravimetric, polarization, and impedance spectroscopies. The performance of extracts was studied for their selection as corrosion inhibitors. The surface of the protected metal was explored using X-ray photoelectron, scanning electron microscopy, and atomic force microscopy. The surface characterization acts as the supportive tolls for the experimental results. Furthermore, theoretical analysis also provides deep insight into the corrosion inhibition at the molecular level, and this is done via analyzing the inhibitor-metal interaction. The extracts contains many heteroatoms that act as adsorption centers, through which inhibitors are adsorbed over the metal surface and protect it from corrosion. The heteroatoms and π-electron clouds, as the center for chemical adsorption and the polar sites, provide physical adsorption. Natural plant extracts are discussed in this chapter to fetter corrosion for a range of metals (steel, alloys, and so on).

Keywords: Plants, corrosion inhibitor, metals, EIS, DFT

4.1 Introduction

In all human activities, metals play a vital role because of their strong mechanical and electrical properties [1–4]. The most common phenomenon that makes a metal weak and fragile is corrosion, which occurs by electrochemical reactions between metals and the aggressive environment [5, 6]. In every part of the industrial application, carbon steel is the most commonly used metal, owing to its superb mechanical abilities. Due to the broader application of carbon steel, it is the most extensive researched. However, copper and aluminum alloys are also well-studied. The loss associated with corrosion can be reduced by the application of corrosion inhibitors [7]. Corrosion inhibitors are added in minimal amounts, and that can mitigate metal corrosion. Presently, the use of synthetic corrosion inhibitors that are not safe to the environment is restricted [8]. Some researchers used expired drugs and plant extracts as the source of corrosion inhibitors [9–13]. The most efficient corrosion inhibitor should be sustainable and friendly to the environment, and that part is well played by plant extracts [14]. Our ability to employ almost every element of a plant

https://doi.org/10.1515/9783110760583-004

as a corrosion inhibitor is a benefit of utilizing plant extract. Figure 4.1 represents the active corrosion inhibitor ingredients present in every part of the plant body.

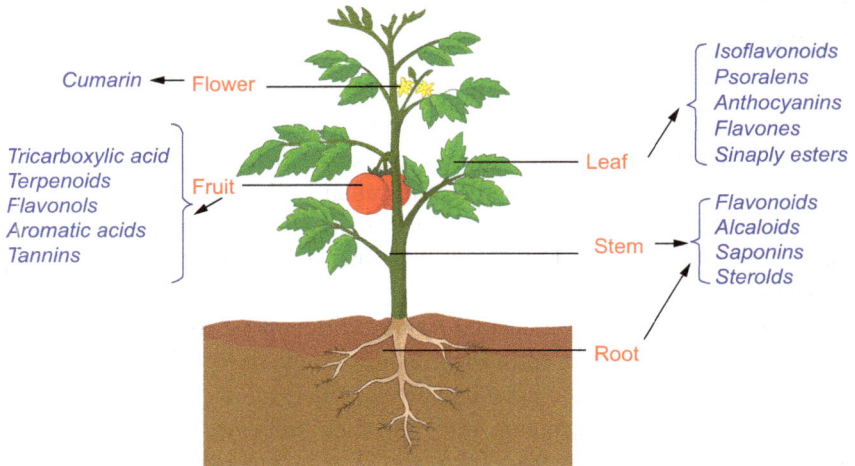

Figure 4.1: Plants parts with standard active components.

Extracts were prepared using almost every part of plants, due to the high concentration of phytochemicals [15, 16]. As per literature, extracts from fruits, seeds, flowers, and leaves have been tested and are found to be efficient corrosion inhibitors due to the active components. Nevertheless, plant parts are very cheap, readily available, and, last but not least, they are a renewable source [17–21]. The present chapter focuses on the application of plant extracts as efficient corrosion inhibitors.

4.2 Carbon steel corrosion inhibitors

Several parts of plants were saturated in the corrosive solution to restrict the corrosion of metals and alloys. Some of them are tabulated in Table 4.1.

Al Hasan et al. analyzed the corrosion inhibition of *Bacopa monnieri* and *Lawsonia inermis* (henna) aqueous extract of the stem [24] and obtained an efficiency of 80% (Table 4.1). Dehghani, Bahlakeh, and Ramazanzadeh obtained leaf extracts of *Eucalyptus* and tested them as a corrosion inhibitor. They obtained a performance of 88% at 800 ppm concentration [28]. Pradipta and co-workers [25] studied the aqueous extract of green tea (*Camellia sinensis*). The inhibitor exhibited an efficiency of 80%. Quraishi et al. [29] extracted the Meethi Neem (*Murraya koenigii*) and used it as a fetter acidic solution. An inhibition efficiency of more than 90% was obtained in both acidic media. Wang and co-workers [31] studied the *Ficus tikoua* leaf extract to impede the

Table 4.1: Details of plant extracts used, metal, corrosive medium, extract concentration, and inhibition efficiency (IE%).

S. no.	Plant extract	Metal/medium	Concentration	IE_{max}%	References
1	*Ananascomosus*	Low-carbon steel/1 M HCl	1,000 mg/L	97.6	[22]
2	*Artemisia herba-alba*	Stainless steel/1 M H_3PO_4	1 g/L	88	[23]
3	*Bacopa monnieri/ Lawsonia inermis*	Low-carbon steel/0.5 M NaOH	0.8 g/L	80.0	[24]
4	*Camellia sinensis*	CS/Sodium Chloride	2%	80	[25]
5	*Cryptocarya nigra*	MS/HCl	1,000 ppm	91	[26]
6	*Sioscorea septemloba*	Carbon steel/HCl	0.1–2.0 g/L	72	[27]
7	*Eucalyptus*	MS/HCl	600 mg/L	88	[28]
8	*Murraya koenigii*	Mild steel/HCl/H_2SO_4	100–600 ppm	95	[29]
9	*Euphorbia heterophylla Linn*	MS/HCl	1 g/L	69	[30]
10	*Ficustikoua*	CS/HCl	200 ppm	95.8	[31]
11	*Glycyrrhiza glabra*	MS/HCl	600 mg/L	88	[32]
12	*Ipomoea batatas*	Galvanized steel/1 M HCl	0.7 g/L	64.26	[33]
13	*Juglansregia*	Mild steel/3.5 wt % NaCl	1,000 ppm	94.2	[34]
14	*Luffa cylindrica*	MS/HCl	1 g/L	87.98	[35]
15	*Nicotiana tabacum*	Q235 steel/ NaOH	34 ppm	83.9	[36]
16	*Olea europaea*	Mild steel/0.1 M NaOH + 0.5 M NaCl	200–800 mg/L	91.9	[37]
17	*Papaver somniferum*	SS/HCl	500 mg/L	88	[38]
18	*Pterocarpussantalinoides*	Low carbon steel/1 mol/dm^3 HCl	0.7 g/L	90	[39]
19	*Rosa canina*	MS/HCl	800 mg/L	86	[40]
20	*Pongamia pinnata*	MS/HCl	400 ppm	97	[41]
21	*Tamarindus indica*	MS/NaCl	1,000 mg/L	96	[42]
22	*Taraxacum officinale*	Carbon steel/seawater	100–400 mg/L	94.3	[43]
23	*Tinospora crispa*	MS/HCl	1,000 mg/L	80	[44]
24	*Tithonia diversifolia*	MS/HCl	0.7% v/v	94.55	[45]
25	*Zingiber officinale*	Mild steel/1 M HCl	100 ppm	92.5	[46]
26	*Ziziphora*	MS/HCl	800 mg/L	93	[47]

corrosion of MS in acidic media. The maximum efficiency was 95.8% at 200 mg/L concentration. SEM images showed a smooth texture with the inhibitor, when compared in its absence. In 2019, Sanaei and group analyzed *Rosa canina* to fetter MS in acid media [40]. The efficiency was explored using EIS, UV–VIS, and surface studies. The maximum corrosion inhibition performance was 86% at 800 ppm. Additionally, the inhibitor is mixed. Singh and co-workers tested the Karanj (*Pongamia pinnata*) aqueous extract over MS in HCl [41]. The results of weight loss and EIS provided an efficiency of 98% and 96%, respectively. The Langmuir isotherm provided the best fit, and the inhibitor is a mixed type. Akbarzadeh et al. [63] analyzed the anticorrosion self-healing coating, based on graphene oxide, Zn^{2+} ions, and *Tamarindus indiaca* (Ti.E) and labelled it as GON-Ti.E-Zn. Numerous techniques, such as Raman spectroscopy, XRD, FT-IR, and UV–VIS were used for the study of inhibitor adsorption. Hussin and co-workers checked *Tinospora crispa* aqueous solution over MS in 1 M HCl [44]. The inhibition results were tested using mass loss, EIS, and polarization (PDP) modes. At 800 ppm concentration, the maximum inhibition was 80%. The Langmuir adsorption isotherm provided the best fitting, and the Gibbs free adsorption energy (ΔGads) was −21.87 kJ/mol. *Taraxacum officinale* extract was used by Deyab and Guibal to study the corrosion of carbon steel in seawater[43]. The corrosion inhibition study was carried out by the mass-loss method, EIS, and PDP. The inhibition came to 94.3%. The PDP result reveals the anodic nature of inhibitor action.

4.3 Studies involving temperature and time interval

The deterioration process is influenced by the change in temperature. Temperature change can modify the adsorption process of active ingredients in the solution. Some extracts lead to amplification of the efficiency by increasing the temperature [48]. However, some tend to decrease the performance with rise in temperature. Therefore, evaluating corrosion inhibitor performance with variation in temperature becomes very important [49–54]. Similarly, the immersion time of metal in the corrosive medium also changes the adsorption chemistry. Thus, this has to be considered in the calculation of the inhibition efficiency [55–58].

Wang et al. group [36] tested the performance of tobacco rob (*Nicotiana tabacum*) in artificial seawater, which contains 0.1 M NaOH, over Q235 steel. In this, nicotine plays a significant role in reducing the oxidization phenomenon. The results of XRD and SEM reveal that in the absence of extract, deposits are observed on the metal. However, in its presence, the expansion of deposits reduced due to the chelation of the −OH group present in the extract. XPS influences the chemical attachment of the extracts' fragments on top of the metal.

Mobin et al. [22] studied the corrosion inhibition property of Pineapple (*Ananas comosus*) stem extracts. The observed efficiency was 97.6% (1,000 mg/L concentration) at 338 K. It is interesting to note that as the temperature increased, the inhibitor performance increased. The inhibitor is a mixed type following the Langmuir adsorption isotherm. At higher temperatures, chemical adsorption was observed. SEM represents the smooth metal surface.

Divya et al. [45] explored the inhibition performance of *Tithonia diversifolia* flower extract over MS in 1 M HCl. The results were tested using weight loss, PDP, and EIS. As the temperature increased, inhibition efficiency increased up to 325 K and attained 94.55%. However, further increase in temperature tended to decrease inhibition efficiency. The results of PDP suggest the mixed nature of extracts.

Sweet potato tuber (PMS) extract was taken out by Anyiam et al. [33]. They evaluated the inhibition on galvanized steel in HCl, at various time and temperatures. The weight loss results give the maximum inhibition efficiency of 64.2% at 0.7 g/L. The corrosion rate increased as the temperature increased. Gadow and Motawea studied the corrosion inhibition effects of ginger roots [46]. The weight loss test provided the value of 94% at 200 mg/L. PDP supports the mixed nature of the inhibitor action. In 2019, Buyuksagis et al. [38] examined the extract of *Papaver somniferum* leaves. The corrosion rate decreased as the temperature rose. The performance of extract at 500 ppm was 88%. AFM, SEM, and energy-dispersive X-ray spectroscopy (EDX) support the thick film of the mitigator on the metal. The adsorption process obeys the Langmuir isotherm, with a mixed nature of inhibition action. Emori group investigated *Dioscorea septemloba* extract as an inhibitor in hydrochloric acid [27]. They observed that an increase in temperature increases the corrosion rate. SEM and FTIR sustain the arrangement of the defensive coating of extracts molecules onto the metal surface. The inhibition efficiency value was 72.1% at 2.0 g/L of inhibitor concentration. Ogunleye group [35] analyzed the inhibition performance of *cylindrica* extract on mild steel in 0.5 M HCl. As temperature increases, the corrosion rate increased; and with increasing inhibitor concentration, corrosion rate decreased. Boudalia et al. studied *Artemisia alba* aqueous solution on SS in 1 M H_3PO_4 solution [23]. They obtained the maximum efficiency value of 88% at 1 g/L and 298 K. The corrosion rate increased as the temperature rose due to the de-lamination of the adsorbed inhibitor film. SEM and EDX support defensive formation onto the metal surface.

Alibakhshi et al. tested the *Glycyrrhiza glabra* root solution as a potential mitigator for MS in acidic medium [59] using PDP and EIS. The results of EIS reveal an inhibition efficiency value of 88% at 800 ppm after 24 h immersion. AFM supports the flat metal face with an inhibitor. Akinbulumo group analyzed the extract of *Euphorbia heterophylla* Linneo on MS in 1.5 M HCl [30]. The weight loss results offer the maximum inhibition efficiency of 69% at 343 K. It was observed that the corrosion rate increases as the temperature increases. Flory–Huggins isotherm offers the best fitting, than Langmuir, El-Awaday, or Temkin. The Gibbs free energy supports the physical nature of inhibitor action. Dehghani group [47] investigated the performance of *Ziziphora* leaf extract over mild steel in 1 M HCl. The value of inhibition is high at higher inhibitor concentration and achieved the value of 93% for 2.5 h immersion. The inhibition efficiency values depend on the immersion time. Gibbs's adsorption-free energy supports the mixed nature of the inhibitor action.

Haddadi and group studied the fruit-shell extract of *Junglans regia* over mild steel in sodium chloride solution [34]. They observed that the inhibitor performance was enhanced when the immersion time increased up to 48 h. At 1,000 ppm, a maximum inhibition efficiency of 94% was obtained. PDP supports that inhibitors influence both anodic and cathodic feedbacks. The adsorption nature of the inhibitor is both physical and chemical.

4.4 Mechanism of adsorption and quantum chemical characterization

The previous sections describe the application of experimental study in the analysis of extracts as corrosion inhibitors. The adsorption mechanism was calculated using Gibbs free adsorption energy (ΔG_{ads}) that is calculated via finding the suitable isotherm. The calculated magnitude of ΔG_{ads} describes the nature of adsorption, such as chemical, physical, or a mixture of both. Table 4.2 describes the extracts of plants, with their important constituents and the applied theoretical calculations.

Table 4.2: Theoretical characterization of the plants' extracts, with computational parameter.

S. no.	Plant extract	Theory	References
1	Dioscorea septemloba	DFT, MD	[27]
2	Eucalyptus	DFT, MC, MD	[28]
3	Ficus tikoua	DFT	[31]
4	Juglans regia	DFT, MC, MD	[34]
5	Glycyrrhiza glabra	DFT, MC, MD	[32]
6	Rosa canina	DFT, MC, MD	[40]
7	Tamarindus indiaca	DFT, MC, MD	[42]
8	Ziziphora	DFT, MC, MD	[47]

Emori et al. analyzed the quantum study for *Dioscorea septemloba* constituents (Table 4.2) [27]. They performed the DFT calculation using RB3LYP/6311++G(d,p) basis sets. The result of energy gap reveals that the performance of glycoside groups is better than the fatty acids group. Furthermore, the results of DFT was supported by molecular dynamics (MD) calculation that estimated the binding energies (E_{bind}), which has the order dioscin > dioscoroneA > sitosterol > palmitic acid. The constituent that has a higher molecular structure has a better inhibition performance. Additionally, π-electrons and heteroatoms undergo back-bonding and make a stronger adsorbed layer over the metal surface.

Dehghani group studied the theoretical calculation of *Eucalyptus* extract. The results of ΔGads supports that the inhibitor undergoes both types of adsorption above the metal plane. Simulations were carried out on both neutral and mono-protonated forms. The DFT results support that electron affluent center-like heteroatoms or aromatic rings donate electron pairs to the empty d-orbitals of iron atoms and facilitate chemical adsorption. Physical adsorption occurs by electrostatic and van der Waals interactions [28].

Quantum chemical study was carried out using DFT, Mote Carlo, molecular dynamic calculations of *Juglans regia* aqueous extract. DFT analysis was performed on the B3LYP/6-311G** basis set [34]. The authors studied both neutral and protonated forms of the major constituent. The results confirmed that the major portion that takes part in the metal-inhibitor interaction includes aromatic rings and heteroatoms [60–62]. The HOMO regions reside over the oxygen heteroatoms and aromatic rings, whereas LUMO presents over the hydroxyl and carbonyl groups.

The major constituents of *Glycyrrhiza glabra* extract were analyzed using DFT on the B3LYP/6-31G** basis set [32]. The effect of water was also considered in the DFT calculation. Monte Carlo and MD were also calculated. The electron pairs present in phenyl rings, methoxy, carbonyl oxygen, and double bond conjugation act as the electron donor to empty d-orbitals. The smaller energy gaps of the constituents make them good electron donor species.

In the same way, Sanaei and group theoretically investigated the corrosion inhibition performance of the extract of *Rosa canina* [40]. They analyzed the adsorption mechanism, in depth, of both neutral and mono-protonated forms of ascorbic acid, marein, pectin, and tannin constituents. The investigation revealed that after protonation, the sites responsible towards adsorption changed [40]. In contrast, the distribution of LUMO was similar, both for neutral and protonated forms. The functional groups, including hydroxyl, carbonyl, and benzene rings, act as the center for inhibitor adsorption.

The active constituents present in the *Ziziphora* aqueous extract [47] were tested by DFT, MC, and MD. The study revealed the flat adsorption of constituents over the metal surface, both for neutral and monoprotonated forms. The heterocyclic rings and heteroatoms are the main sites that donate electrons to the metal surface. The MD results well support the DFT calculation. The authors suggest a positively charged metal surface and thus, Cl^- ions can be easily adsorbed, which promote the adsorption of the protonated *Ziziphora* leaf extract molecules via electrostatic interaction.

4.5 Corrosion inhibitors for other metals

The water and methanol extract of *Borassus flabellifer* was tested as a corrosion inhibitor for aluminum in 1 M H_2SO_4 solution [64]. The result of the potentiodynamic polarization (PDP) and electrochemical impedance spectroscopy (EIS) revealed that inhibition efficiency increases with increasing extract concentration. The result of PDP suggests the maximum protection performance of 66.88% (methanol) and 51.85% (water) at 0.40 g/L concentrations (Table 4.3). Also, PDP suggests that the inhibitor is mixed in nature. SEM pictures support the formation of an inhibitor protective layer over the aluminum surface.

Al-Nami and Fouda studied *Commiphora myrrha* as corrosion mitigator for copper in a 2 M HNO_3 [65]. The authors tested the inhibitor performance using weight loss (WL), PDP, electrochemical frequency (EFM), and EIS. All tested methods revealed that as the inhibitor concentration increased, corrosion inhibition performance increased, and reached 91.8% at 300 mg/L. The formation of an inhibitor protective film was confirmed by FT-IR, AFM, and SEM techniques. The PDP curve reveals the mixed nature of inhibitor action.

Table 4.3: Plant extracts as copper and aluminum corrosion inhibitors.

S. no.	Plant extract	Metal/medium	Concentration	IE$_{max}$%	References
1	*Borassus flabellifer*	Al/1 M H_2SO_4	0.1–0.4 g/L	66.8	[64]
2	*Commiphora myrrha*	Cu/2 M HNO_3	50–300 ppm	91	[65]
3	*Equisetum arvense*	Cu/Seawater	250–1,000 ppm	87.5	[66]
4	*Hemerocallis fulva*	Al/1 M H_2SO_4	200–600 ppm	89	[67]

Esquivel-Lopez et al. studied the *Equisetum arvense* extract for copper in sodium chloride [66]. The EIS result suggests that an increasing inhibitor amount increases the inhibition power, and the maximum value is 87.5%. The chemical structure of the active constituent was confirmed by gas chromatography, mass spectrometry, and FT-IR. SEM images confirmed the formation of an inhibitor layer over the metal surface.

The corrosion protection of aluminum in 1 M H_2SO_4 was carried out by *Hemerocallis fulva* methanol extract [67]. The obtained results suggest that maximum protection efficiency is 89% at 600 mg/L and 303 K. Langmuir provided the best fitting of the experimental data with physical nature adsorption. PDP reveals the mixed nature of inhibitor action on corrosion inhibition. SEM, EDS, and AFM support the inhibitor adsorption and formation of the protective layer.

The above study suggests that plant extracts are a better source as corrosion mitigators. Additionally, they are acquired from sustainable resources and are nonhazardous/green compounds [68, 69]. Plants' extract consist of different active constituents that include heterocyclic rings and they provide a clue to synthetic organic scientist to develop these types of compounds that are also synthetic in nature and are beneficial to the industry and environment.

4.6 Conclusion

The chapter provided a short review of the application of plant extracts as corrosion inhibitors on various metallic surfaces. The evaluation of plant extracts as corrosion inhibitor was carried out by varying their concentration, solvent of extraction, temperature, etc. using different techniques, such as WL, EIS, PDP, EFM, etc. The adsorption model, including the Langmuir isotherm, was tested, which describes the nature of the extract adsorption – physical, chemical, or a mixture of both. Physical adsorption mainly occurs via the Polar Regions that are present in the inhibitor molecules. The electron sharing by electron-rich centers, such as heteroatoms and π-electron clouds over phenyl ring to vacant metal orbitals, are responsible for chemical adsorption. The extracts consist of various phytochemicals that include organic compounds with heteroatoms that undergo inhibitor-metal interaction. The theoretical study, including DFT, MC, and MD, helps in predicting and supporting the experiment findings and also provides the molecular orientation of the active constituents. Some results of the plant extracts as corrosion inhibitors for copper and aluminum were also explained. Although plant extracts are green toward the environment, the isolation of active components is a very challenging task.

References

[1] Parthipan P, Elumalai P, Narenkumar J, Machuca LL, Murugan K, Karthikeyan OP, Rajasekar A. *Allium sativum* (garlic extract) as a green corrosion inhibitor with biocidal properties for the control of MIC in carbon steel and stainless steel in oilfield environments. International Biodeterioration & Biodegradation, 2018; 132: 66–73.
[2] Loto RT, Olowoyo O. Synergistic effect of sage and jojoba oil extracts on the corrosion inhibition of mild steel in dilute acid solution. Procedia Manufacturing, 2019; 35: 310–314.
[3] Anupama KK, Ramya K, Joseph A. Electrochemical measurements and theoretical calculations on the inhibitive interaction of *Plectranthus amboinicus* leaf extract with mild steel in hydrochloric acid. Measurement, 2017; 95: 297–305.
[4] Singh A, Ebenso EE, Quraishi MA. Corrosion inhibition of carbon steel in HCl solution by some plant extracts. International Journal of Corrosion, 2012; Article ID 897430. 10.1155/2012/897430.
[5] Singh A, Ebenso EE, Quraishi MA. *Boerhavia diffusa* (Punarnava) root extract as green corrosion inhibitor for mild steel in hydrochloric acid solution: Theoretical and electrochemical studies. International Journal of Electrochemical Science, 2012; 7: 8659–8675.

[6] Singh A, Ebenso EE, Quraishi MA. Theoretical and electrochemical studies of *Cuminum Cyminum* (Jeera) extract as green corrosion inhibitor for mild steel in hydrochloric acid solution. International Journal of Electrochemical Science, 2012; 7: 8543–8559.

[7] Singh A, Ahamad I, Singh VK, Quraishi MA. The effect of environmentally benign fruit extract of Shahjan (*Moringa oleifera*) on the corrosion of mild steel in hydrochloric acid solution. Chemical Engineering Communications, 2012; 199: 63–77.

[8] Singh A, Ebenso EE, Quraishi MA. Stem extract of brahmi (*Bacopa monnieri*) as green corrosion inhibitor for aluminum in NaOH solution. International Journal of Electrochemical Science, 2012; 7: 3409–3419.

[9] Singh A, Ahamad I, Singh VK, Quraishi MA. *Piper longum* (Pepper) extract as green corrosion inhibitor for aluminum in NaOH solution. Arabian Journal of Chemistry, 2016; 9: S1584–S1589.

[10] El-Haddad MN, Fouda AS, Hassan AF. Data from chemical, electrochemical and quantum chemical studies for interaction between Cephapirin drug as an eco-friendly corrosion inhibitor and carbon steel surface in acidic medium. Chemical Data Collection, 2019; 22: 100251.

[11] Singh A, Lin Y, Liu W, Kuanhai D, Pan J, Huang B, Ren C, Zeng D. A study on the inhibition of N80 steel in 3.5% NaCl solution saturated with CO2 by fruit extract of Gingko biloba. Journal of the Taiwan Institute of Chemical Engineers, 2014; 45: 1918–1926.

[12] Farahati R, Mousavi-Khoshdel SM, Ghaffarinejad A, Behzadi H. Experimental and computational study of penicillamine drug and cysteine as water-soluble green corrosion inhibitors of mild steel. Progress in Organic Coating, 2020; 142: 105567.

[13] Espinoza-Vázquez A, Rodríguez-Gómez FJ, Negrón-Silva GE, González-Olvera R, Ángeles-Beltrán D, Palomar-Pardavé M, Miralrio A, Castro M. Fluconazole and fragments as corrosion inhibitors of API 5LX52 steel immersed in 1 M HCl. Corrosion Science, 2020; 174: 108853.

[14] Olawale O, Bello JO, Ogunsemi BT, Uchella UC, Oluyori AP, Oladejo NK. Optimization of chicken nail extracts as corrosion inhibitor on mild steel in 2 M H2SO4. Heliyon, 2019; 5: e02821.

[15] Pradeep Kumar CB, Mohana KN. Phytochemical screening and corrosion inhibitive behavior of *Pterolobium hexapetalum* and *Celosia argentea* plant extracts on mild steel in industrial water medium. Egyptian Journal of Petroleum, 2014; 23: 201–211.

[16] Marsoul A, Ijjaali M, Elhajjaji F, Taleb M, Salim R, Boukir A. Phytochemical screening, total phenolic and flavonoid methanolic extract of pomegranate bark (Punica granatum L): Evaluation of the inhibitory effect in acidic medium 1 M HCl. Material Today Proceeding, 2020. 27: 3193–3198.

[17] Sedik A, Lerari D, Salci A, Athmani S, Bachari K, Gecibesler ˙IH, Solmaz R. Dardagan fruit extractas eco-friendly corrosion inhibitor for mild steel in 1 M HCl: Electrochemical and surface morphological studies. Journal of Taiwan Institute of Chemical Engineer, 2020; 107: 189–200.

[18] Dehghani A, Bahlakeh G, Ramezanzadeh B, Ramezanzadeh M. Potential of Borage flower aqueous extract as an environmentally sustainable corrosion inhibitor for acid corrosion of mild steel: Electrochemical and theoretical studies. Journal of Molecular Liquids, 2019; 277: 895–911.

[19] Singh A, Lin Y, Liu W, Yu S, Pan J, Ren C, Kuanhai D. Plant derived cationic dye (Berberine) as an effective corrosion inhibitor for 7075 aluminium alloy in 3.5% NaCl solution. Journal of Industrial and Engineering Chemistry, 2014; 20: 4276–4285.

[20] Singh A, Lin Y, Ebenso EE, Liu W, Pan J, Huang B. *Gingko biloba* fruit extract as an eco-friendly corrosion inhibitor for J55 steel in CO$_2$ saturated 3.5% NaCl solution. Journal of Industrial and Engineering Chemistry, 2015; 24: 219–228.

[21] Singh A, Lin Y, Ebenso EE, Liu W, Kuanhai D, Pan J, Huang B. Relevance of electrochemical and surface studies to probe *Zanthoxylum schinifolium* (sichuan pepper) as an effective corrosion inhibitor for N80 steel in CO2 saturated 3.5% NaCl solution. International Journal of Electrochemical Science, 2014; 9: 5585–5595.

[22] Mobin M, Basik M, Aslam J. Pineapple stem extract (*Bromelain*) as an environmental friendly novel corrosion inhibitor for low carbon steel in 1 M HCl. Measurement, 2019; 134: 595–605.

[23] Boudalia M, Fernández-Domene RM, Tabyaoui M, Bellaouchou A, Guenbour A, García-Antón J. Green approach to corrosion inhibition of stainless steel in phosphoric acid of Artemesia herba alba medium using plant extract. Journal of Materials Research and Technology, 2019; 8: 5763–5773.

[24] Al Hasan NHJ, Alaradi HJ, Al Mansor ZAK, Al Shadood AHJ. The dual effect of stem extract of Brahmi (Bacopa monnieri) and Henna as a green corrosion inhibitor for low carbon steel in 0.5 M NaOH solution. Case Studies in Construction Materials, 2019; 11: e00300.

[25] Pradipta I, Kong D, Tan JBL. Natural organic antioxidants from green tea inhibit corrosion of steel reinforcing bars embedded in mortar. Construction and Building Materials, 2019; 227: 117058.

[26] Faiz M, Zahari A, Awang K, Hussin H. Corrosion inhibition on mild steel in 1 M HCl solution by *Cryptocarya nigra* extracts and three of its constituents (alkaloids). RSC Advances, 2020; 10: 6547–6562.

[27] Emori W, Zhang R-H, Okafor PC, Zheng X-W, He T, Wei K, Lin X-Z, Cheng C-R. Adsorption and corrosion inhibition performance of multi-phyto constituents from Dioscorea septemloba on carbon steel in acidic media: Characterization, experimental and theoretical studies. Colloids and Surfaces. A, Physicochemical and Engineering Aspects, 2020; 590: 124534.

[28] Dehghani A, Bahlakeh G, Ramezanzadeh B. Green Eucalyptus leaf extract: A potent source of bio-active corrosion inhibitors for mild steel. Bioelectrochemistry, 2019; 130: 107339.

[29] Quraishi MA, Singh A, Singh VK, Yadav DK, Singh AK. Green approach to corrosion inhibition of mild steel in hydrochloric acid and sulphuric acid solutions by the extract of Meethi Neem (*Murraya koenigii*) leaves. Materials Chemistry and Physics, 2010; 122: 114–122.

[30] Akinbulumo OA, Odejobi OJ, Odekanle EL. Thermodynamics and adsorption study of the corrosion inhibition of mild steel by *Euphorbia heterophylla* L. extract in 1.5 M HCl. Results in Materials, 2020; 5: 100074.

[31] Wang Q, Tan B, Bao H, Xie Y, Mou Y, Li P, Chen D, Shi Y, Li X, Yang W. Evaluation of *Ficus tikoua* leaves extract as an eco-friendly corrosion inhibitor for carbon steel in HCl media. Bioelectrochemistry, 2019; 128: 49–55.

[32] Alibakhshi E, Ramezanzadeh M, Haddadi SA, Bahlakeh G, Ramezanzadeh B, Mahdavian M. *Persian Liquorice* extract as a highly efficient sustainable corrosion inhibitor for mild steel in sodium chloride solution. Journal of Cleaner Production, 2019; 210: 660–672.

[33] Anyiam CK, Ogbobe O, Oguzie EE, Madufor IC, Nwanonenyi SC, Onuegbu GC, Obasi HC, Chidiebere MA. Corrosion inhibition of galvanized steel in hydrochloric acid medium by a physically modified starch. SN Applied Sciences, 2020; 2: 520.

[34] Haddadi SA, Alibakhshi E, Bahlakeh G, Ramezanzadeh B, Mahdavian M. A detailed atomic level computational and electrochemical exploration of the *Juglans regia* green fruit shell extract as a sustainable and highly efficient green corrosion inhibitor for mild steel in 3.5 wt % NaCl solution. Journal of Molecular Liquids, 2019; 284: 682–699.

[35] Ogunleye OO, Arinkoola AO, Eletta OA, Agbede OO, Osho YA, Morakinyo AF, Hamed JO. Green corrosion inhibition and adsorption characteristics of Luffa *cylindrica* leaf extract on mild steel in hydrochloric acid environment. Heliyon, 2020; 6: e03205.

[36] Wang H, Gao M, Guo Y, Yang Y, Hu R. A natural extract of tobacco rob as scale and corrosion inhibitor in artificial seawater. Desalination, 2016; 398: 198–207.

[37] Ben Harb M, Abubshait S, Etteyeb N, Kamoun M, Dhouib A. Olive leaf extract as a green corrosion inhibitor of reinforced concrete contaminated with seawater. Arabian Journal of Chemistry, 2020; 13: 4846–4856.

[38] Buyuksagis A, D'Ilek M. The use of *Papaver somniferum* L. Plant extract as corrosion inhibitor. Protection of Metals and Physical Chemistry of Surfaces, 2019; 55: 1182–1194.

[39] Ahanotu CC, Onyeachu IB, Solomon MM, Chikwe IS, Chikwe OB, Eziukwu CA. *Pterocarpussantalinoides* leaves extract as a sustainable and potent inhibitor for low carbon steel in a simulated pickling medium. Sustainable Chemistry and Pharmacy, 2020; 15: 100196.

[40] Sanaei Z, Ramezanzadeh M, Bahlakeh G, Ramezanzadeh B. Use of *Rosa canina* fruit extract as a green corrosion inhibitor for mild steel in 1 M HCl solution: A complementary experimental, molecular dynamicsand quantum mechanics investigation. Journal of Industrial and Engineering Chemistry, 2019; 69: 18–31.

[41] Singh A, Ahamad I, Singh VK, Quraishi MA. Inhibition effect of environmentally benign Karanj (*Pongamia pinnata*) seed extract on corrosion of mild steel in hydrochloric acid solution. Journal of Solid State Electrochemistry, 2011; 15: 1087–1097.

[42] Akbarzadeh S, Ramezanzadeh B, Bahlakeh G, Ramezanzadeh M. Molecular/electronic/ atomic-level simulation and experimental exploration of the corrosion inhibiting molecules attraction at the steel/chloride-containing solution interface. Journal of Molecular Liquids, 2019; 296: 111809.

[43] Deyab MA, Guibal E. Enhancement of corrosion resistance of the cooling systems in desalination plants by green inhibitor. Scientific Report, 2020; 10: 4812.

[44] Hussin MH, Jain Kassim M, Razali NN, Dahon NH, Nasshorudin D. The extract of *Tinospora crispa* as a natural mild steel corrosion inhibitor in 1 M HCl solution. Arabian Journal Chemistry, 2016; 9: S616–S624.

[45] Divya P, Subhashini S, Prithiba A, Rajalakshmi R. *Tithonia diversifolia* flower extract as green corrosion inhibitor for mild steel in acid medium. Material Today Proceeding, 2019; 18: 1581–1591.

[46] Gadow HS, Motawea MM. Investigation of the corrosion inhibition of carbon steel in hydrochloric acid solution by using ginger roots extract. RSC Advances, 2017; 7: 24576–24588.

[47] Dehghani A, Bahlakeh G, Ramezanzadeh B, Ramezanzadeh M. Potential role of a novel green eco-friendly inhibitor in corrosion inhibition of mild steel in HCl solution: Detailed macro/ micro-scale experimental and computational explorations. Construction and Building Materials, 2020; 245: 118464.

[48] Bahlakeh G, Dehghani A, Ramezanzadeh B, Ramezanzadeh M. Highly effective mild steel corrosion inhibition in 1 M HCl solution by novel green aqueous Mustard seed extract: Experimental, electronic-scale DFT and atomic-scale MC/MD explorations. Journal of Molecular Liquids, 2019; 293: 111559.

[49] Anupama KK, Ramya K, Shainy KM, Joseph A. Adsorption and electrochemical studies of *Pimenta dioica* leaf extracts as corrosion inhibitor for mild steel in hydrochloric acid. Materials Chemistry and Physics, 2015; 167: 28–41.

[50] Oguzie EE, Enenebeaku CK, Akalezi CO, Okoro SC, Ayuk AA, Ejike EN. Adsorption and corrosion-inhibiting effect of *Dacryodis edulis* extract on low-carbon-steel corrosion in acidic media. Journal of Colloid and Interface Science, 2010; 349: 283–292.

[51] Mourya P, Banerjee S, Singh MM. Corrosion inhibition of mild steel in acidic solution by *Tagetes erecta* (Marigold flower) extract as a green inhibitor. Corrosion Science, 2014; 85: 352–363.

[52] Bouknana D, Hammouti B, Messali M, Aouniti A, Sbaa M. *Olive pomace* extract (OPE) as corrosion inhibitor for steel in HCl medium. Asian Pacific Journal of Tropical Disease, 2014; 4: S963–S974.

[53] Muthukrishnan P, Jeyaprabha B, Prakash P. Adsorption and corrosion inhibiting behavior of *Lannea coromandelica* leaf extract on mild steel corrosion. Arabian Journal of Chemistry, 2017; 10: S2343–S2354.

[54] Hamdy A, El-Gendy NS. Thermodynamic, adsorption and electrochemical studies for corrosion inhibition of carbon steel by henna extract in acid medium. Egyptian Journal of Petroleum, 2013; 22: 17–25.

[55] Dehghani A, Bahlakeh G, Ramezanzadeh B. A detailed electrochemical/theoretical exploration of the aqueous Chinese gooseberry fruit shell extract as a green and cheap corrosion inhibitor for mild steel in acidic solution. Journal of Molecular Liquids, 2019; 282: 366–384.

[56] Gerengi H, Uygur I, Solomon M, Yildiz M, Goksu H. Evaluation of the inhibitive effect of *Diospyros kaki* (Persimmon) leaves extract on St37 steel corrosion in acid medium. Sustainable Chemistry and Pharmacy, 2016; 4: 57–66.

[57] Muthukrishnan P, Prakash P, Jeyaprabha B, Shankar K. Stigmasterol extracted from *Ficus hispida* leaves as a green inhibitor for the mild steel corrosion in 1 M HCl solution. Arabian Journal of Chemistry, 2019; 12: 3345–3356.

[58] Kumar KPV, Pillai MSN, Thusnavis GR. Seed extract of *Psidium guajava* as ecofriendly corrosion inhibitor for carbon steel in hydrochloric acid medium. Journal of Materials Science & Technology, 2011; 27: 1143–1149.

[59] Alibakhshi E, Ramezanzadeh M, Bahlakeh G, Ramezanzadeh B, Mahdavian M, Motamedi M. *Glycyrrhiza glabra* leaves extract as a green corrosion inhibitor for mild steel in 1 M hydrochloric acid solution: Experimental, molecular dynamics, Monte Carlo and quantum mechanics study. Journal of Molecular Liquids, 2018; 255: 185–198.

[60] Zhang J, Qiao G, Hu S, Yan Y, Ren Z, Yu L. Theoretical evaluation of corrosion inhibition performance of imidazoline compounds with different hydrophilic groups. Corrosion Science, 2011; 53: 147–152.

[61] Khaled KF. Studies of iron corrosion inhibition using chemical, electrochemical and computer simulation techniques. Electrochemical Acta, 2010; 55: 6523–6532.

[62] Musa AY, Jalgham RTT, Mohamad AB. Molecular dynamic and quantum chemical calculations for phthalazine derivatives as corrosion inhibitors of mild steel in 1 M HCl. Corrosion Science, 2012; 56: 176–183.

[63] Akbarzadeh S, Ramezanzadeh M, Ramezanzadeh B, Bahlakeh G. A green assisted route for the fabrication of a high-efficiency self-healing anticorrosion coating through graphene oxide nanoplatform reduction by Tamarindus indiaca extract. Journal of Hazardous Materials, 2020; 390: 122147.

[64] Nathiya RS, Perumal S, Murugesan V, Raj V. Evaluation of extracts of *Borassus flabellifer* dust as green inhibitors for aluminium corrosion in acidic media. Materials Science in Semiconductor Processing, 2019; 104: 104674.

[65] Al-Nami S. Corrosion inhibition effect and adsorption activities of methanolic myrrh extract for Cu in 2 MHNO3. International Journal of Electrochemical Science, 2020; 15: 1187–1205.

[66] Esquivel López A, Cuevas-Arteaga C, Valladares-Cisneros MG. Universidad Autónoma del Estado deMorelos study of the corrosion inhibition of copper in synthetic seawater by *Equisetum arvense* as green corrosion inhibitor. Revista Mexicana de Ingeniería Química, 2019; 19: 603–616.

[67] Chung I-M, Malathy R, Kim S-H, Kalaiselvi K, Prabakaran M, Gopiraman M. Eco-friendly green inhibitor from *Hemerocallis fulva* against aluminum corrosion in sulphuric acid medium. Journal of Adhesion Science and Technology, 2020; 34: 1483–1506.

[68] Sastri VS. *Green Corrosion Inhibitors: Theory and Practice*, John Wiley and Sons: Hoboken, NJ, USA, 2011. ISBN 1-118-01417-0.

[69] Pedraza Basulto GK, Carrillo I, Ortega D, Martinez L, Canto J. Evaluation at pipeline corrosion at oil field. ECS Transactions, 2015; 64: 103–110.

Omotayo Sanni, Jianwei Ren, Tien-Chien Jen

5 Biomass waste as corrosion inhibitor for metals in corrosive media

Abstract: Due to increasing pollution that can be attributed primarily to fossil fuel consumption the world has been facing unprecedented challenges. This has caused global warming and aggravated the current pandemic (COVID-19) problem, as well. Owing to the increasing demand for fossil fuels and excessive consumption of petroleum-based resources, it has become vital to adopt alternative renewable sources. In recent times, biomass waste is being used efficiently as an ideal green corrosion inhibitor because of its availability, ecological acceptability, biodegradability, renewability, phytoconstituent-rich nature, and, most significantly, nontoxic nature. Their valorizations expand potential application in the industry rather than "waste to energy" in the circular economy perception. There is a strong interest globally in developing suitable technology that can use biomass wastes for different applications, including corrosion inhibitors. In this regard, this chapter presents a collection of articles designed to inform metallurgists, designers, and engineers of the nature of corrosion and means of its prevention. Beginning with the analysis of fundamental scientific principles involved in corrosion science, it covers the basics as well as other topics such as causes of corrosion, methods of mitigating corrosion, the use and characteristics of inhibitors, classes of corrosion inhibitors, factors to consider in selecting corrosion inhibitor, and classification of corrosion inhibitors. Attention is given to biomass waste inhibitors through literature review and the biomass wastes extract used as corrosion inhibitors and their acting mechanisms are presented. Sections include an introduction, corrosion inhibitors, and biomass waste extract as corrosion inhibitors.

Keywords: Biomass waste, corrosion inhibitors, metals, corrosive media

5.1 Introduction

Pure alloys and metals react electrochemically/chemically with an aggressive environment to form stable compounds in which corrosion occurs. Corrosion involves metal ions movement into the solutions at the active area (anodes), electron passage from the metal to acceptor at a less active area (cathodes), electronic current in the metal, and an ionic current in the solution. The cathodic process requires the presence of electron acceptors such as hydrogen ions, oxygen, or oxidizing agent [1]. In general, corrosion of metal can simply be described as the deterioration of metallic

https://doi.org/10.1515/9783110760583-005

properties in contact with some elements present in the environment, which is, normally, an unavoidable process.

Corrosion is known to be detrimental to human well-being and the environment. Corrosion is a damaging phenomenon affecting different industries – for instance, mechanical and textiles industries, leading to chemical leakage and rupture of the corroded pipelines in the oil and gas industry [2–4]. As summarized in Figure 5.1, the cause of corrosion includes acid or bases, water present in the air, aggressive metal polish, hazardous gases, salts, and liquid chemicals that can instigate corrosion on the surface of the metal [5–7]. Generally, alloys and metals mainly show a high propensity towards corrosion owing to the acid present in the environment. For example, acid solutions are used for industrial applications such as acid descaling, acid pickling, acid cleaning, and mill scale removal from the surface of the metal. In the oil and gas sector, corrosion can be linked to the nature of crude oil that promotes corrosion because of its destructive impurities like naphthenic and sulfuric acid [8–10].

There are two types of corrosion, namely, wet and dry corrosion. Wet corrosion takes place in the presence of liquid that has an electrolyte. Chloride solution, acids, and seawater are typical electrolyte media where this type of corrosion takes place. This form of metal corrosion takes place via electron transfer, which involves oxidation and reduction processes. The surrounding environments gain electrons in reduction, while the metal atoms lose electrons in oxidation. In contrast, dry corrosion takes place without water or moisture to support the degradation process. This form of corrosion happens because of direct attacks on the surface of the metal by atmospheric gases such as anhydrous inorganic liquids, oxygen, sulfur dioxide, halogen, hydrogen sulfide, and nitrogen in the environment. These processes are temperature-sensitive and take place in high-temperature systems. In wet corrosion, the corrosion rate is much faster than in dry corrosion, since water acts as electrolytes in wet corrosion and, therefore, aids the electron movement from the anode to cathode. Different research groups have extensively reported the corrosion process in wet conditions [11–19]. Inadequate corrosion prevention results in corrosion-related cost issues according to literature, for instance, replacement of damaged structures, maintenance, rehabilitation, and repair [20–22]. Recently, studies conducted by NACE in 2016 suggest that corrosion causes loss of around 2.5 trillion US dollars in the economy, which represents almost 3.4 percent of the total Gross Domestic Product. This has encouraged researchers to focus on corrosion as a significant issue that needs attention, globally. These studies reveal that by implementing existing corrosion prevention technology properly, the corrosion cost can be minimized by 15–35% (375–875 billion US dollars). Due to safety and extremely high economic loss concerns, corrosion is a significant subject that must be tackled by engineers and scientists working in the field, worldwide. Corrosion prevention is more achievable and practical rather than total elimination. Several practical methods to control or prevent corrosion in the manufacture, chemical, fertilizer, mineral, energy, and food industry include:

- Application of coatings, linings, and paints for protecting industrial energy and plants
- Selection of corrosion-resistant material
- Cathodic protection using sacrificial anodic metal, for instance, magnesium and aluminum alloy
- Corrosion inhibition

Among these methods, the use of corrosion inhibitor is an excellent approach in avoiding the devastating destruction of alloys and metals in corrosive environments [23–29].

Corrosion inhibition is an accepted technique; the use of corrosion inhibitors in the industry is increasing globally in different industrial and technological applications, such as cleaning and pickling steel vehicles during production, natural gas and petroleum oil facilities, transportation, steel-reinforced concrete structures, military equipment, and storage of electronics. Application of corrosion inhibitor is an efficient and viable technique in controlling the rate of corrosion, because this material can be applied easily via continuous and/or batch treatments with a small quantity of material, with the aim of minimizing complete or partial plant shutdowns.

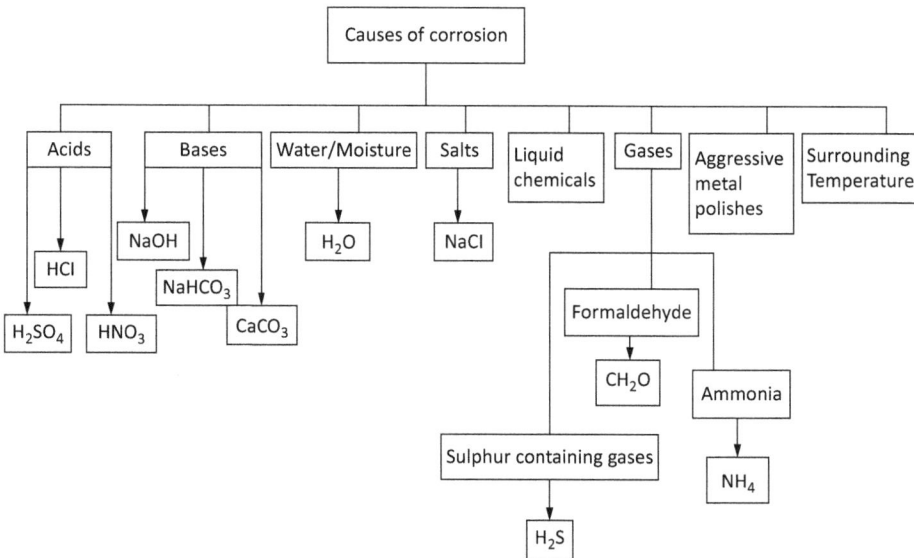

Figure 5.1: Causes of metal and alloy corrosion. Adapted with permission from [131] (Copyright (2021) Elsevier).

5.2 Corrosion inhibitors

Generally, corrosion inhibitors can simply be described as substances, which, when added in small concentrations to an aggressive solution, can efficiently decrease the rate of metal corrosion [30, 31]. The inhibitor can be added to several systems: chemicals, refinery, cooling systems, boiler, oil and gas production unit, etc. The inhibitor works via adsorption of ion or molecule onto the metal surface and the adsorption of inhibitor molecule is classified based on their inhibition effect (Figure 5.2).

Inhibitors reduce the rate of corrosion by:
- Decreasing the diffusion rate from the reactant to the metal surface.
- Decreasing or increasing the cathodic and/or anodic reaction.
- Decreasing the metallic surface electrical resistance.

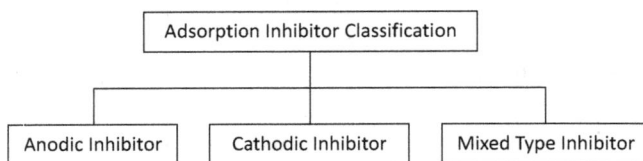

Figure 5.2: Classification of inhibitor adsorption, based on inhibition effect (adapted with permission from [132]. Copyright Reza N A, Akhmal N H, Fadil N A, Taib M F M, some rights reserved; exclusive licensee [MPDI]. Distributed under a Creative Commons Attribution License 4.0 (CC BY) https://creativecommons.org/licenses/by/4.0/).

The application of inhibitors is easy and offers the benefit of in situ applications, without significantly interrupting the process. However, there are different factors that are to be considered when choosing an appropriate inhibitor:
- Environmental friendliness
- Toxicity of inhibitors
- Cost of inhibitors
- Quantity of inhibitors
- Availability of inhibitors

An inhibitor can be adsorbed onto the metal surface via:
- Electrostatic attraction between the charged molecule of the inhibitor and the surface of the metal charged (electrostatic adsorption).
- The π-bond orbital is adsorbed by the interaction of the conjugated bond in the inhibitor with the metal surface.
- Interaction between the unshared pair electron molecules with the surface of the metal.
- Combination of points 1 and 2.

The inhibitor films adsorbed act as a barrier between the metal surface and the surrounding environments, thus protecting the metal surface from corroding. In general, organic inhibitor heteroatoms exist in a polar functional group, for instance, $-CN$, $-OCH_3$, $-NO_2$, $-OH$, $-COOH$, $-NH_2$, or $-COOC_2H_5$, which acts as adsorption center during adsorption of this compound onto the metal surface. Different types of green corrosion inhibitors have been reported in inhibiting metal corrosion effectively; these groups or classes are summarized in Figure 5.3. Inorganic, organic compounds, and mixtures of the organic-inorganic compounds have been employed successfully in different aggressive media for most alloys and metals. Although several synthetic compounds demonstrated excellent anticorrosive performance, unfortunately, many of them are very toxic to the environment and humans. The environmental and safety issues of corrosion inhibitors in the industry have been a global concern. Inhibitors may cause temporary or permanent damage to organ systems, viz., liver or kidneys. The toxicity might manifest either during its applications or during the synthesis of the compound. The application of most corrosion inhibitors is limited because of their hazardous effect on human health, for instance, chromium-based treatment [32, 33]. This toxic effect has led to the use of natural products as an anticorrosion agent, which is harmless and biodegradable. In recent times, strict environmental protocols and growing ecological awareness, worldwide, have prevented the use and synthesis of toxic traditional inhibitors. Therefore, the application of conventional inhibitors is now restricted, due to the increasing idea of "green chemistry" in the field of science and engineering. Green chemistry uses a set of principles that cause reduction in the discharge of malignant materials into the surrounding environment. Lately, strict eco-friendly regulations and increasing ecological consciousness in the science and engineering fields have encouraged engineers and scientists in the corrosion science field to move in the direction of synthesizing organic inhibitors by one-step multicomponent reactions that combine three or more reacting molecules in a single step. Therefore, the application of conventional corrosion inhibitors is now limited due to the increasing "green chemistry" concept in the field of engineering, technology, and science, in order to reduce the negative impact of toxic compounds by selecting low toxic chemicals.

Eliminating the inherent danger of specific processes or products, green chemistry can generate better environmental benefits, thus moving them beyond many environmental standards. Nowadays, finding environmentally friendly and inexpensive inhibitors is one of the main goals of researchers. Different natural products such as *Opuntia Ficus-Indica* [34–42], Ginkgo leaf [43–45], Nettle leaves [46–49], barley agroindustrial waste [50, 51], *Pisum sativum* peels [52], *Allium sativum* (garlic extract) [53–56], grape pomace [57–59], *Borago officinalis* L. [60, 61], red algae *Halopitys incurvus* [62], *Rollinia occidentalis* [63], *Cuscuta reflexa* [64], *Matricaria recutita chamomile* [65, 66], *Prosopis juliflora* [67, 68], *Artemisia judaica* herbs [69], *Glycyrrhiza glabra* leaves [70, 71], *Raphanus sativus* L. [72, 73], *Crataegus oxyacantha* [74], *Prunus avium* plant [75], mature areca nut husk [76], *Sida cordifolia* [77], and *Pongamia pinnata* [78–80] have been reported as possible corrosion inhibitors. The increased interest in green inhibitors has encouraged several scientists/engineers to investigate natural plant-based inhibitors, globally. Nevertheless, the extracted inhibitors from wastes with little or no harm to the environment and plants are still scarce. In the current ecological context, food, plants, and agro-industrial waste extracts emerge as alternatives in fulfilling the condition of the REACH regulation and European directives on wastewater rejection. The inhibitive tendency of agro-food wastes and plant extract is usually ascribed to the phytochemical compounds naturally present in this substance, which has antioxidant properties [81].

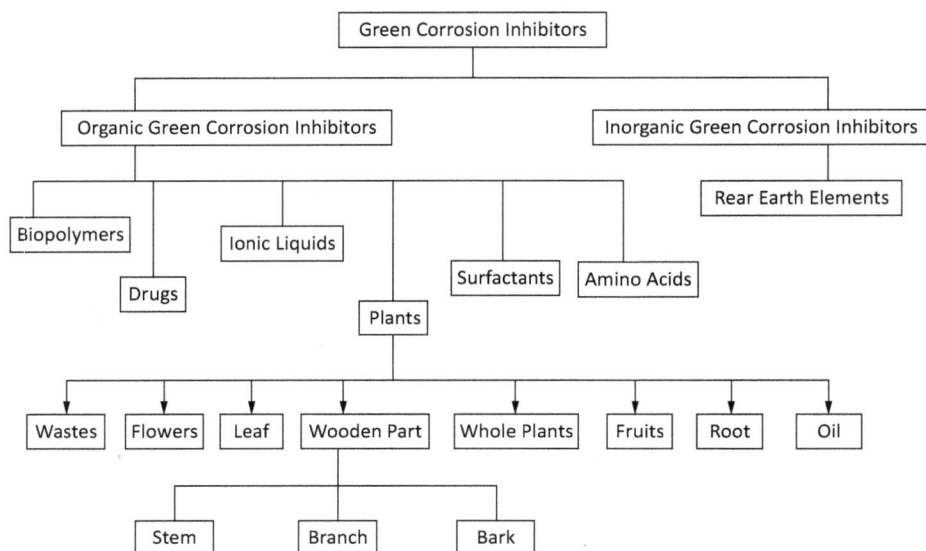

Figure 5.3: Groups of green corrosion inhibitors (adapted with permission from [131], Copyright (2021) Elsevier).

5.3 Biomass wastes

The world has been facing exceptional challenges owing to the increased pollution, which can be attributed primarily to fossil fuel consumption. This has caused global warming and aggravated the recent pandemic (COVID-19) problem [82, 83]. This challenge has made researchers search for raw materials from natural sources. In this chapter, we focus on biomass wastes. Green chemistry has attracted vast attention in the last decade, in several contexts, with commercial products and chemical technology aiming to minimize wastes and reduce toxins. There has been a rising concern since the 1980s, about the use of hazardous and toxic chemicals and the generation of wastes. The motivation for the turning point in environmental awareness was given by the USA announcement on the Pollution and Prevention Act, in 1990. Environmental concerns and their solution is now a global priority. Preventing wastes from forming at the origin, with source reduction program and other related principles was publicized by the "12 Principles of Green Chemistry" [84]. This principle was intended as a design and guide standard for molecular scientists.

Green chemistry has demonstrated how fundamental scientific methodology can protect human health and the environment in an economical way, by integrating areas such as catalysis, synthetic and analytical method development, solvents, and designing safe chemicals with an increased awareness for impacts on the environment. In this perspective, biomass material is at the base of the alignment of several industrial appliances to the principle of sustainability and green chemistry. Different forms of biomass are necessary for animal nutrition and humans; this means that using biomass for other reasons must be balanced. The remaining biomass after removing the nutritious component, the biowastes, ideally serves as feedstock for many applications that could be related to green chemistry. Major areas where green chemistry is frequently applied to biomass-derived products are the innovations for minimizing wastes and environmental impacts relating to metal surface protection. The lignocellulose biomass produced globally every year has been reported [85, 86]. The little amount of the biomass generated is used for medicinal plants, extracting wood, non-edible purposes, edible fruits, crops, etc. In most scenarios, agricultural remains are burnt on site, thereby causing immense difficulties for people not only onsite but also in far away places, leading to air pollution, and as a result, contributing to global warming. It has been predicted that around one-third of all food produced nowadays goes to landfills. However, wastes from diverse sources present an opportunity to use biomass wastes. They can be used as corrosion inhibitors that are eco-friendly and biodegradable.

There is a strong growing interest globally, in developing an appropriate technology that can obtain materials and chemicals from biomass in open literature for different applications, including corrosion inhibitors. Plant extract with S, O, and N heteroatoms have good adsorptive properties leading to excellent anticorrosive properties [87, 88]. The plant extract is prepared from the plant material via different extraction and solvents methods. As a result, an organic molecule extracted from food by-products appears as a substitute in the corrosion inhibition field, owing to their availability and biodegradability. Therefore, highly efficient, environment-friendly corrosion inhibitors derived from natural products such as diverse parts of the plant – fruits, roots, leaves, stems, seeds, and flowers – attract high consideration, of late, by different research groups. Reports on diverse bio-based inhibitors show that phytochemicals can serve as effective corrosion inhibitors [89, 90]. Metallic corrosion is a significant issue in the industry, and has, therefore, found productive research in the field of green chemistry. Green inhibitors are receiving huge attention in the corrosion field owing to their recyclability, renewability, and ecological acceptance. This includes alkaloids [91], agricultural-wastes [92, 93], amino acids [94], plants extracts, and polyphenols [95]. Their valorizations expand possible application in industries, rather than waste to energy. This chapter is a combined outline of significant works in the open literature, with a focus on the use of biomass waste as a green corrosion inhibitor.

5.3.1 Biomass wastes extract as corrosion inhibitors

In the twenty-first century, biomass energy has the ability to become one of the promising resources, because of its advantages such as wide distribution, low pollution, and renewability. Biomass extract is usually preferred as a green corrosion inhibitor, because of its potential nature, biodegradability, availability, medicinal properties, phyto-constituent-rich and highly non toxic nature. This chapter investigates the potential tendency of biomass waste products in moving towards green chemistry. Among the wide variety of biomass, syringaldehyde and vanillin as typical lignin derivatives, are used extensively for preparing polyesters, composites, phenolic resins, etc. Novel polymers based on syringaldehyde and vanillin have been developed recently [96]. Meanwhile, functional polymers bearing aldehyde groups can work as amine dosimeters, enzyme carriers, and adsorbents. Additionally, syringaldehyde-based polymers have better thermal stability than vanillin.

Presently, there are many studies on biomass application in the field of corrosion inhibitors. Garlic powder was examined by Yaro et al. [97] as an acidic corrosion inhibitor via combinations of experiments and theoretical computations. Fadhil et al. [98] investigated the inhibitive tendency of Portulaca grandiflora extract on mild steel corrosion in acid solutions. Sulaiman et al. [99] examined the behavior of quercetin, bonducellin, and citrulline as green compounds. The order of inhibition was quercetin > bonducellin > citrulline, based on the obtained results. Parthipan et al. [100] studied the inhibition efficiency of Glycyrrhiza glabra extracts as an eco-friendly inhibitor for corrosion of steel in 1 molar hydrochloric acid. The results from the experiment show that the inhibitor acts as mixed-type, with optimum efficiency of 88% at 800 ppm. Naseri et al. [101] investigated the inhibitive tendency of Clopidogrel as a non-toxic inhibitor for mild steel in acidic solutions. The adsorption tests showed that Clopidogrel was adsorbed chemically and physically on the metal surface. The majority of reported corrosion inhibitors are effective at low temperatures. With an increase in temperature to around 80 °C, the efficiency of the inhibitors decreases; this inhibitor was reported to maintain over 90% efficiency at high temperature. The utilization of (N1-(2-aminoethyl)-N2-(2-(2-(furan-2-yl)-4, 5-dihydro-1H-imidazol-1-yl) ethyl ethane-1,2-diamine) (NNED) lingo cellulose renewable resources in plant-based biomass in an acid environment using the electrochemical and gravimetrical analysis was reported by Zhan Chen et al. [102]. Surface morphology methods such as SEM/EDS, FT-IR, and AFM were used to investigate the inhibiting layer structure on the steel surface. Also, the quantum chemical calculation was used as a theoretical tool to understand the mechanisms of inhibition. The authors reported inhibitive performance greater than 90% at an inhibitor concentration of 5 ppm. The adsorption of the inhibitor on the surface of the metal occurs spontaneously and fits the Langmuir adsorption isotherms.

The effect of *S. surattense* plant with sharp thorn, which belongs to the Solanaceae family also known as Thai green eggplant, frequently found all around Egypt, India, Nepal, China, Bangladesh, and Bhutan has reported [103, 104] good anticorrosion properties in this case. Aloe vera extract [105, 106], Rosa canina fruit extract [107], and Persian licorice extracts [108] were examined as corrosion inhibitors for different materials in chloride and acid solutions. Disposing walnut green husk as an agricultural waste from walnut production is a perplexing issue due to its harmful influence on the atmosphere, water, and soil. In recent times, walnut green husk extract is used as an inhibitor, additive, and natural dye. The corrosion of cold-rolled steel in 0.10 M trichloroacetic acid solutions using potassium iodide and walnut green husk has been reported by Liu et al. [109]. In this regard, 97.2% efficiency of the inhibitor was reported. The synergic effect of walnut green husk and sodium lignosulfonate for cold-rolled steel in phosphoric acid solutions was studied, and 97.2% inhibition efficiency was reported [110]. It has also been reported for magnesium alloys in the NaCl solutions. The extract of elephant grass as a nontoxic, cheap, renewable, and readily available blend with halide ions was reported by Ituen et al. [111]. The authors reported that the electron donor's site like N, and π-electrons, and O generally regarded as potential adsorption sites in other efficient corrosion inhibitors, were found in phyto-compounds like lignin, flavonoids, and anthocyanins. Aldehyde-containing microspheres (ACMs) starting from syringaldehyde have received attention. Onion peels regarded as food by-products that are normally discarded as wastes have received attention lately, because they are easy to source locally. In this regard, processing onion peels into oilfield chemicals will not compete with the food but will create wealth out of wastes. Also, since onion is mostly consumed in households worldwide, it is a sustainably available feedstock. The phytochemical analysis of onion shows the presence of flavonoids, anthraquinones saponins, terpenoids, tannins, cardiac glycosides, and terpenoids [112]. The high-performance liquid chromatographic analysis of the extract shows that the extract is rich in quercetin, ferulic acid, protocatechuic acid, kaempferol, and gallic acid. Red onion peel extract was reported to inhibit the corrosion of aluminum and steel in different media [113]. The temperature range was varied to simulate harsh surface situations encountered in different fields. Natural honey has been reported as corrosion inhibitor for different materials in aggressive solutions [114, 115]. The anticorrosion activity of bitter gourd, onion, and garlic for mild steel in hydrochloric solutions has been reported to show excellent results [116]. Inhibiting mild steel against corrosion using ethanolic extracts of *Ricimus communis* leaves in acid media has been reported [117]. The aqueous extract of *Agaricus* and hibiscus flower was examined as corrosion inhibitor for industrial cooling systems by Minhaj et al. [118]. Pure organic compounds extracted from natural products such as amino acids, ascorbic acid, caffeine, and succinic acid are also reported as corrosion inhibitors. The anticorrosion effects of tea wastes [119], guar gum [120], extract of *Rosmarinus officinalis* [121], and *Andrographis paniculata* [122] have been reported. The inhibitive effects of an organic derivative of plant extract, artesunate (dihydroartemisinin-12α-succinate; C19H28O8), a semi-synthetic derivative of artemisinin, which is a Chinese herb extract of *Artemisia annua*, and rutin (quercein-3-rutinoside; C27H30O16), a flavonol glycoside made up of disaccharide rutinose and flavonol quercetin, which is a major constituent of extract from orange peels have been reported. Rutin is also found in apple peel, black tea, and buckwheat. Interestingly, the corrosion inhibitive effect of crude aqueous extract from orange peels has been previously reported [123]. The authors reported that the corrosion rate and constant rate decreased with increasing surface coverage, while the half-life increased. The values of ΔG_{ads} calculated indicate spontaneous physisorption process, and the adsorption characteristic fits the Langmuir isotherm. Algae extracts are used widely in the aesthetic and medical fields. Microalgae have substantial biotechnological potentials, and their biomass can be used in the production of biofuels, bio fertilizers, food, animal feed, and bioactive compounds. Nevertheless, a few reports are available in open literature about their uses as corrosion inhibitors. The biomass effects from microalgae *Chlorella sorokiniana* for mild steel corrosion in hydrochloride solutions using potentiodynamic polarization technique, weight-loss, surface morphological, and electrochemical impedance spectroscopy analysis was investigated [124], and a good result was achieved in this regard.

An inhibitive efficiency of 94.6% after 24 h of immersion was achieved using the *Chlorella sorokiniana* biomass. Microalgae are eukaryotic or prokaryotic microorganisms that perform aerobic photosynthesis. They have an extensive range of natural composites such as vitamins, pigments, polysaccharides, phenolic compounds, proteins, lipids, and fatty acids. They are used in different industries like bioenergy applications, pharmaceutical, food, and cosmetic industries. There are about 35 different spirulina strains, including *S. maxima*. This microalga is known for its excellent amount of protein, 50–70% in its composition. A different study suggests a composition between 13–25% carbohydrates and 5–7% of total lipids in the composition of these microalgae. It also has more than 12 amino acids, minerals such as sodium, calcium, iron, magnesium, manganese, zinc, potassium, vitamins (A, B1, B2, B6, B12, E), and phosphorus. The microalgae, *Spirulina maxima*, were examined as natural corrosion inhibitors for carbon steel in acid solutions by electrochemical impedance measurements, potentiodynamic polarization curves, surface, and gravimetric analysis. After 72 h of immersion, 96.4% was achieved for the biomass of the microalgae *Spirulina maxima* [125]. Castor oil as a biodegradable source was reported as a good corrosion inhibitor. Sunflower oil as biomass and renewable source for the development of new green corrosion inhibitor that can show high efficiency in aggressive acid environments at high temperatures was reported [126]. The authors reported that sunflower oil could provide an inhibition efficiency of 98% and 93% at 60 and 80 °C, respectively. The inhibiting effect of fruit extracts of Terminalia chebula on the corrosion of mild steel in 1 M HCl solutions in broadening applications of biomass extracts for metallic corrosion was reported by Singh et al. [127]. The authors establish the effectiveness of biomass extracts as potential corrosion inhibitors for practical applications. Caffeic acid, derived from lignin, has also been investigated as a corrosion inhibitor. The efficiency of coffee extracts and tea as green inhibitors was studied; the highest inhibition efficiency of 79.02% at 1.5 g/L tea extract inhibitor solution with 12 days of immersion time was achieved [128]. The inhibitive performance of ethanol extract of bell pepper (*Capsicum frutescens*) on low carbon steel corrosion in acid solutions has been investigated using computational and experimental techniques. Since ancient times, fruits of bell pepper-like pepper fruits have been used in traditional medicine and food flavor.

The pectin derived from citrus peel has been reported as a corrosion inhibitor. Chitosan is extracted from seafood wastes; fresh leaves of watermelon, banana, and sugarcane; seed and peel of watermelons; by-products of tomato [129, 130]. The by-products of tomato seeds, peels, and pulps, used as animal feed or disposed of, usually, as solid waste have been used as a pectin source for inhibiting corrosion of tin. Different authors were using food wastes without complete awareness of the significance of the used matrix. The corrosion-inhibitive performance of most plant extracts thus decreases at high temperatures due to the deprivation of their complex phytochemicals. This restricts their use in the field of corrosion inhibition, as recovered hydrocarbon from many rewarding pay zones requires high-temperature systems. It would be advantageous if corrosion inhibitors from plants are optimized to be efficient at elevated temperatures. Findings from this study will draw the attention of academics and industrialists, particularly surface and corrosion engineers, materials scientists, oilfield chemists, and corrosion inhibitors manufacturers.

5.3.2 Mechanism of inhibition

According to Bockris and Drazic [133], the inhibition mechanism can be explained following equation:

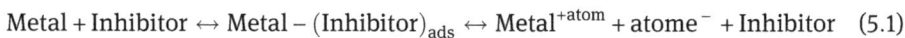

$$\text{Metal} + \text{Inhibitor} \leftrightarrow \text{Metal} - (\text{Inhibitor})_{ads} \leftrightarrow \text{Metal}^{+atom} + \text{atome}^- + \text{Inhibitor} \quad (5.1)$$

The Metal-(Inhibitor)$_{ads}$ intermediates form an adsorption film via the atom of the inhibitor. The adsorption film acts as a hindrance to the solution and enhances the protection of the metal surface. The adsorption of an organic compound on the surface of the metal electrode is regarded as a substitutional adsorption process between the organic compound in the aqueous phase (Org$_{ad}$) and the water molecules adsorbed on the metal surface (H$_2$O$_{ad}$) according to the following equation:

$$Org_{aqueous} + XH_2O_{adsorbed} \leftrightarrow Org_{adsorbed} + H_2O_{aqeous} \qquad (5.2)$$

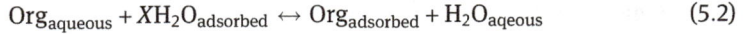

where X is the number of water molecules replaced by one organic molecule.

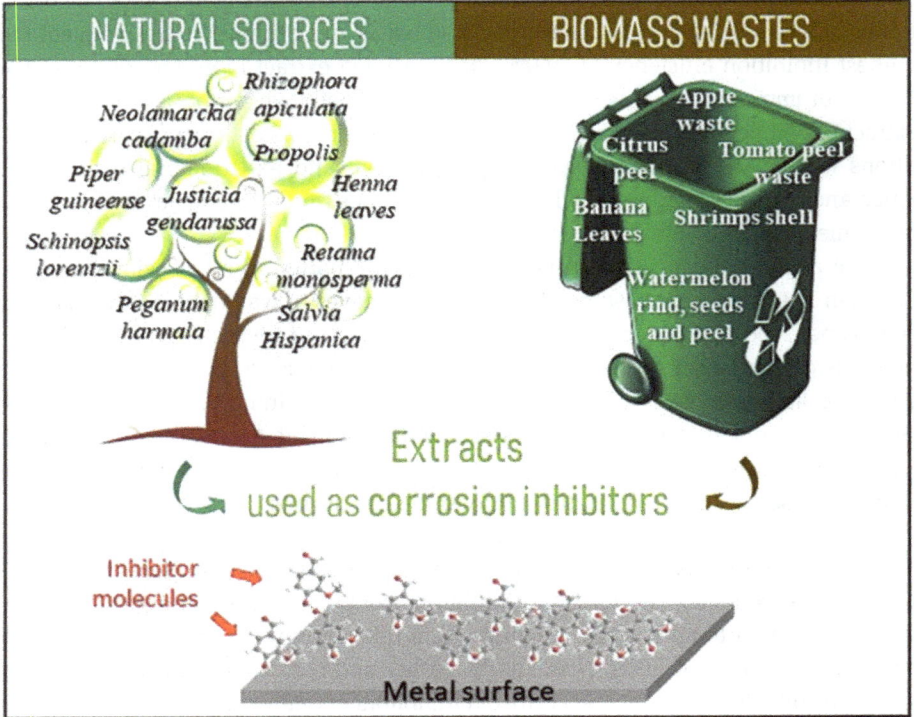

Figure 5.4: Illustration of biomass waste mechanism of inhibition (adapted with permission from [133]. Copyright Marzorati S, Verotta L, Trasatti S P, some rights reserved; exclusive licensee [MPDI]. Distributed under a Creative Commons Attribution License 4.0 (CC BY) https://creativecommons.org/licenses/by/4.0/).

The inhibitive effect of biomass waste is owing to the adsorption of the biomass waste compound on the metal surface as shown in Figure 5.4, thereby blocking the active site. Two steps occur, primarily, in the process of corrosion inhibition: transfer of inhibitor on the surface of the metal followed by combinations of phytochemical functional groups with the metal surface. Also, the mechanism of biomass waste inhibition depends on the functional and phytochemical groups in the biomass wastes.

5.3.3 Future scope

Recently, there has been extensive interest in organic green corrosion inhibitors from extracts of biomass wastes such as peels and seeds. This is due to the high amount of biomass waste produced annually. The study on biomass waste is far behind the research conducted on plant-based organic green corrosion inhibitors, despite the increase in interest. Also, the various potentials of the field are yet to be investigated, and the validation of the existing results is essential to use this option as a scalable industrial solution. Some of the research directions that can further substantiate the corrosion inhibition potential of biomass wastes are as follows: detailed characterization of the corrosion inhibition molecules is crucial to form a foundation and thorough scale-up study in the industrial context. The adaption of computer-aided tools to optimize the experimental data on biomass waste inhibitors could also be integrated to warrant sustainable technology as an industrially adaptable choice.

5.4 Conclusion

Corrosion is a global challenge that needs to be promptly and properly tackled. The range of research results on corrosion control using biomass waste inhibitors summarized in this chapter advocate the prospects of bioinspired agents as excellent corrosion inhibitors that can replace toxic, inorganic, synthetic, and traditional expensive corrosion inhibitors, because they are nonhazardous, readily available, safe, and biosustainable with natural environments and can replace toxic compounds causing severe hazard to the environments. The biomass extracts contain several constituents/ phytochemicals that can be easily adsorbed and inhibit metal against deterioration. Phytochemicals from biomass extract, typically heterocyclic compound, with a heteroatom such as P, N, S, and O adsorbed efficiently on the metal surface and form defensive films, acting as a physical barrier in restricting the molecules or ions diffusion on the metal surface and, consequently, retard the corrosion reaction rate. The biomass extracts as corrosion inhibitors are usually examined via electrochemical means, which includes measurements like electrochemical impedance spectroscopy, potentiodynamic polarization, surface studies for instance SEM, UV, and AFM, and weight loss. Most of the extracts adsorbed on the metal surface fit the Langmuir adsorption isotherms, though a few report Freundlich and Temkin adsorption isotherms: via chemisorption, physisorption, or mixed mechanisms.

References

[1]   Tamalmani K, Husin H. Review on corrosion inhibitors for oil and gas corrosion issues. Applied Sciences, 2020; 10: 3389.

[2]   Heidersbach R. Metallurgy and Corrosion Control in Oil and Gas Production. John Wiley & Sons, 2018, 1–4.

[3]   Wood MH, Arellano AV, Van Wijk L Corrosion related accidents in petroleum refineries. European Commission Joint Research Centre, report no. EUR, 2013, 26331.

[4]   Ossai CI Corrosion related accidents in petroleum refineries. European Commission Joint Research Centre. International Scholarly Research Notices, 2012, 2012.

[5]   Raja PB, Ismail M, Ghoreishiamiri S, Mirza J, Ismail MC, Kakooei S, Rahim AA. Reviews on corrosion inhibitors: A short view. Chemical Engineering Communications, 2016; 203: 1145–1156.

[6]   Marrone PA, Hong GT. Corrosion control methods in supercritical water oxidation and gasification processes. Paper presented at the CORROSION 2008, New Orleans, Louisiana, March 2008.

[7]   Salleh SZ, Yusoff AH, Zakaria SK, Taib MAA, Seman AA, Masri MN, Ter Teo P. Plant extracts as green corrosion inhibitor for ferrous metal alloys: A review. Journal of Cleaner Production, 2021; 304: 127030.

[8]   Mostafaei A, Peighambari SM, Nasirpouri F. Failure analysis of monel packing in atmospheric distillation tower under the service in the presence of corrosive gases. Engineering Failure Analysis, 2013; 28: 241–251.

[9]   Bhowmik PK, Hossain M, Shamim JA Corrosion and its control in crude oil refining process in 6th International Mechanical Engineering & 14th Conference Annual Paper Meet (6IMEC&14APM), Dhaka, Bangladesh, 2012.

[10]   Groysman A. Corrosion Problems and Solutions in Oil Refining and Petrochemical Industry. Gewerbestrasse, Switzerland: Springer International Publishing, 2017, Vol. 32, 1–7.

[11]   Koochaki MS, Khorasani SN, Neisiany RE, Ashrafi A, Trasatti SP, Magni M. A highly responsive healing agent for the autonomous repair of anti-corrosion coatings on wet surfaces. In operando assessment of the self-healing process. Journal of Materials Science, 2021; 56: 1794–1813.

[12]   Kawasaki Y, Tomoda Y, Ohtsu M. AE monitoring of corrosion process in cyclic wet–dry test. Construction and Building Materials, 2010; 24: 2353–2357.

[13]   Gong K, Wu M, Xie F, Liu G, Sun D. Effect of dry/wet ratio and pH on the stress corrosion cracking behavior of rusted X100 steel in an alternating dry/wet environment. Construction and Building Materials, 2020; 260: 120478.

[14]   Qian B, Hou B, Zheng M. The inhibition effect of tannic acid on mild steel corrosion in seawater wet/dry cyclic conditions. Corrosion Science, 2013; 72: 1–9.

[15]   Wu J, Cui J, Zheng Q, Zhang S, Sun W, Yang B, Wang JQ. Insight into the corrosion evolution of Fe-based amorphous coatings under wet-dry cyclic conditions. Electrochimica Acta, 2019; 319: 966–980.

[16]   Hu S, Liu L, Cui Y, Li Y, Wang F. Influence of hydrostatic pressure on the corrosion behavior of 90/10 copper-nickel alloy tube under alternating dry and wet condition. Corrosion Science, 2019; 146: 202–212.

[17]   Gong K, Wu M, Xie F, Liu G. Effect of dissolved oxygen concentration on stress corrosion cracking behavior of pre-corroded X100 steel base metal and the welded joint in wet–dry cycle conditions. Journal of Natural Gas Science and Engineering, 2020; 77: 103264.

[18] Wang L, Xing Y, Liu Z, Zhang D, Du C, Li X. Erosion–corrosion behavior of 2205 duplex stainless steel in wet gas environments. Journal of Natural Gas Science and Engineering, 2016; 35: 928–934.

[19] Gong K, Wu M, Liu G. Comparative study on corrosion behaviour of rusted X100 steel in dry/wet cycle and immersion environments. Construction and Building Materials, 2020; 235: 117440.

[20] Koch G. Cost of corrosion. Trends in Oil and Gas Corrosion Research and Technologies, 2017; 3–30. https://doi.org/10.1016/B978-0-08-101105-8.00001-2

[21] Roberge PR. Handbook of Corrosion Engineering. McGraw-Hill Education, New York 2019.

[22] Carsana M, Gastaldi M, Redaelli E. A case study on corrosion conditions and guidelines for repair of a reinforced concrete chimney in industrial environment. Structure and Infrastructure Engineering, 2021; 1–12. doi:10.1080/15732479.2021.1945115

[23] Palanisamy G. Corrosion Inhibitors. Intechopen: London, 2019.

[24] Sanni O, Fayomi OSI, Popoola API. Eco-friendly inhibitors for corrosion protection of stainless steel: An overview. Journal of Physics. Conference Series, 2019; 1374: 042047.

[25] Dwivedi A, Bharti P, Shukla SK. An overview of the polymeric materials that can be used to prevent metal corrosion: A review. Journal of the Turkish Chemical Society Section A: Chemistry, 2021; 8: 863–872.

[26] Brycki B, Szulc A. Gemini surfactants as corrosion inhibitors. A review. Journal of Molecular Liquids, 2021; 344: 117686.

[27] Zehra S, Mobin M, Aslam J. An overview of the corrosion chemistry. Environmentally Sustainable Corrosion Inhibitors, 2022; 3–23. doi.org/10.1016/B978-0-323-85405-4.00012-4

[28] Sayin K, Karakaş D. Quantum chemical studies on the some inorganic corrosion inhibitors. Corrosion Science, 2013; 77: 37–45.

[29] Khadom AA, Kadhim MM, Anaee RA, Mahood HB, Mahdi MS, Salman AW. Theoretical evaluation of Citrus Aurantium leaf extract as green inhibitor for chemical and biological corrosion of mild steel in acidic solution: Statistical, molecular dynamics, docking, and quantum mechanics study. Journal of Molecular Liquids, 2021; 343: 116978.

[30] Saeed MT, Saleem M, Usmani S, Malik IA, Al-Shammari F, Deen KM. Corrosion inhibition of mild steel in 1 M HCl by sweet melon peel extract. Journal of King Saud University-Science, 2019; 31: 1344–1351.

[31] Meriem Z, Hana F, Souad D, Abderrazak B, Amin MA, Leila R, Belakhdar A, Jeon BH, Boulechfar C, Benguerba Y. Experimental and theoretical evaluation of the adsorption process of some polyphenols and their corrosion inhibitory properties on mild steel in acidic media. Journal of Environmental Chemical Engineering, 2021; 9: 106482.

[32] El-Enin SA, Amin A. Review of corrosion inhibitors for industrial applications. International Journal of Engineering Research and Reviews, 2020; 3: 127–145.

[33] Izionworu V, Ukpaka C, Oguzie E. Green and eco-benign corrosion inhibition agents: Alternatives and options to chemical based toxic corrosion inhibitors. Chemistry International, 2020; 6: 232–259.

[34] Torres-Acosta AA. Opuntia-Ficus-Indica (Nopal) mucilage as a steel corrosion inhibitor in alkaline media. Journal of Applied Electrochemistry, 2007; 37: 835–841.

[35] Flores-De los Rios J, Sanchez-Carrillo M, Nava-Dino C, Chacon-Nava J, Gonzalez-Rodríguez J, Huape-Padilla E, Martinez-Villafane A. Opuntia ficus-indica extract as green corrosion inhibitor for carbon steel in 1 M HCl solution. Journal of Spectroscopy, 2015; 2015: 1–9.

[36] Ghazi Z, ELmssellem H, Ramdani M, Chetouani A, Rmil R, Aouniti A, Hammounti B. Corrosion inhibition by naturally occurring substance containing Opuntia-Ficus Indica extract on the corrosion of steel in hydrochloric acid. Journal of Chemical and Pharmaceutical Research, 2014; 6: 1417–1425.

[37] Krid F, Zouaoui E, Medjram MS. Aqueous extracts of Opuntia Ficus-Indica as green corrosion inhibitor of A283C carbon steel IN 1N sulfuric acid solution. Chemistry & Chemical Technology, 2018; 12: 405–409.

[38] Suarez-Hernandez R, Gonzalez-Rodriguez JG, Dominguez-Patiño GF, Martinez-Villafañe A. Use of Opuntia ficus extract as a corrosion inhibitor for carbon steel in acidic media. Anti-Corrosion Methods and Materials, 2014; 61: 224–231.

[39] Galea IE The use of Opuntia ficus-indica as a corrosion inhibitor in steel reinforced concrete. Master's thesis, University of Malta, 2018.

[40] Torres-Acosta AA, González-Calderón PY. Opuntia Ficus-Indica (OFI) Mucilage as corrosion inhibitor of steel in CO_2-contaminated mortar. Materials, 2021; 14: 1316.

[41] Boulmerka R, Abderrahmane S. Opuntia ficus-indica as a durable and eco-friendly corrosion inhibitor on AISI 410 stainless steel in 0.5 MH_2SO_4. Bulletin of Materials Science, 2020; 43: 1–16.

[42] Saidi N, Elmsellem H, Ramdani M, Yousfi F, Rmili R, Azzaoui K, Chahboun N. A Moroccan Opuntia Ficus Indica methanolic flowers extract as an eco-friendly antioxidant and anti-corrosion for mild steel in 1 M HCl. Journal of Material and Environmental Science, 2016; 7: 4105–4115.

[43] Qiang Y, Zhang S, Tan B, Chen S. Evaluation of Ginkgo leaf extract as an eco-friendly corrosion inhibitor of X70 steel in HCl solution. Corrosion Science, 2018; 133: 6–16.

[44] Singh A, Lin Y, Ebenso EE, Liu W, Pan J, Huang B. Gingko biloba fruit extract as an eco-friendly corrosion inhibitor for J55 steel in CO_2 saturated 3.5% NaCl solution. Journal of Industrial and Engineering Chemistry, 2015; 24: 219–228.

[45] Deng S, Li X. Inhibition by Ginkgo leaves extract of the corrosion of steel in HCl and H_2SO_4 solutions. Corrosion Science, 2012; 55: 407–415.

[46] Ramezanzadeh M, Bahlakeh G, Ramezanzadeh B, Sanaei Z. Adsorption mechanism and synergistic corrosion-inhibiting effect between the green Nettle leaves extract and Zn^{2+} cations on carbon steel. Journal of Industrial and Engineering Chemistry, 2019; 77: 323–343.

[47] Ramezanzadeh M, Sanaei Z, Bahlakeh G, Ramezanzadeh B. Highly effective inhibition of mild steel corrosion in 3.5% NaCl solution by green Nettle leaves extract and synergistic effect of eco-friendly cerium nitrate additive: Experimental, MD simulation and QM investigations. Journal of Molecular Liquids, 2018; 256: 67–83.

[48] Nasibi M, Mohammady M, Ghasemi E, Ashrafi A, Zaarei D, Rashed G. Corrosion inhibition of mild steel by Nettle (Urtica dioica L.) extract: Polarization, EIS, AFM, SEM and EDS studies. Journal of Adhesion Science and Technology, 2013; 27: 1873–1885.

[49] Zaabar A, Aitout R, Makhloufi L, Belhamel K, Saidani B. Inhibition of acid corrosion of mild steel by aqueous nettle extracts. Pigment & Resin Technology, 2014; 43: 127–138.

[50] Matos LAC, Taborda MC, Alves GJT, Cunha M, Banczek Oliveira EM, Rodrigues PRP. Application of an acid extract of barley agro-industrial waste as a corrosion inhibitor for stainless steel AISI 304 in H $_2SO_4$. International Journal of Electrochemical Science, 2018; 13: 1577–1593.

[51] Santos ÉdCd, Cordeiro R, Santos Md, Rodrigues PRP, Singh A, D'Elia E. Barley agro-industrial residues as corrosion inhibitor for mild steel in 1mol L-1HCl solution. Materials Research, 2019; 22: 1–14.

[52] Srivastava M, Tiwari P, Srivastava S, Kumar A, Ji G, Prakash R. Low cost aqueous extract of Pisum sativum peels for inhibition of mild steel corrosion. Journal of Molecular Liquids, 2018; 254: 357–368.

[53] Parthipan P, Elumalai P, Narenkumar J, Machuca LL, Murugan K, Karthikeyan OP, Rajasekar A. Allium sativum (garlic extract) as a green corrosion inhibitor with biocidal properties for

the control of MIC in carbon steel and stainless steel in oilfield environments. International Bio deterioration & Biodegradation, 2018; 132: 66–73.

[54]   Hajsafari N, Razaghi Z, Tabaian SH. Electrochemical study and molecular dynamics (MD) simulation of aluminum in the presence of garlic extract as a green inhibitor. Journal of Molecular Liquids, 2021; 336: 116386.

[55]   Barreto LS, Tokumoto MS, Guedes ICd, Melo HG, Amado FDR, Capelossi VR. Evaluation of the anticorrosion performance of peel garlic extract as corrosion inhibitor for ASTM 1020 carbon steel in acidic solution. Matéria (Rio de Janeiro), 2017; 22: 1–14.

[56]   Asfia MP, Rezaei M, Bahlakeh G. Corrosion prevention of AISI 304 stainless steel in hydrochloric acid medium using garlic extract as a green corrosion inhibitor: Electrochemical and theoretical studies. Journal of Molecular Liquids, 2020; 315: 113679.

[57]   Vorobyova V, Chygyrynets O, Skiba M, Zhuk T, Kurmakova I, Bondar O. A comprehensive study of grape pomace extract and its active components as effective vapour phase corrosion inhibitor of mild steel. International Journal of Corrosion and Scale Inhibition, 2018; 7: 185–202.

[58]   Vorobyova V, Skiba M. A novel eco-friendly vapor phase corrosion inhibitor of mild steel. Pigment & Resin Technology, 2019; 48: 137–147.

[59]   Vorobyova V, Skiba M. Apricot pomace extract as a natural corrosion inhibitor of mild steel corrosion in 0.5 m NaCl solution: A combined experimental and theoretical approach. Journal of Chemical Technology & Metallurgy, 2020; 55: 210–222.

[60]   Al-Moubaraki AH. Potential of borage flowers aqueous extract, Borago officinalis L., against the corrosion of mild steel in phosphoric acid. Anti-Corrosion Methods and Materials, 2018; 65: 53–65.

[61]   Draouet C, Hamaidia K, Brakni A, Boutemedjet S, Soltani N. Ethanolic extracts of Borago officinalis L. affect growth, development and energy reserve profile in the mosquito Culex pipiens. Journal of Entomological Research, 2020; 44: 203–210.

[62]   Benabbouha T, Siniti M, El Attari H, Chefira K, Chibi F, Nmila R, Rchid H. Red Algae Halopitys Incurvus extract as a green corrosion inhibitor of carbon steel in hydrochloric acid. Journal of Bio- and Tribo-Corrosion, 2018; 4: 1–9.

[63]   Alvarez PE, Fiori-Bimbi MV, Neske A, Brandan SA, Gervasi CA. Rollinia occidentalis extract as green corrosion inhibitor for carbon steel in HCl solution. Journal of Industrial and Engineering Chemistry, 2018; 58: 92–99.

[64]   Saxena A, Prasad D, Haldhar R. Investigation of corrosion inhibition effect and adsorption activities of Cuscuta reflexa extract for mild steel in 0.5 M H_2SO_4. Bioelectrochemistry, 2018; 124: 156–164.

[65]   Nasr K, Fedel M, Essalah K, Deflorian F, Souissi N. Experimental and theoretical study of Matricaria recutita chamomile extract as corrosion inhibitor for steel in neutral chloride media. Anti-Corrosion Methods and Materials, 2018; 65: 292–309.

[66]   Hamrahi B, Fattah-alhosseini A, Gashti SO, Khanlarkhani A, Madani SM. Green corrosion inhibitor of the Matricaria Chamomilla oil to protect steel in the acidic H2S-containing solution. Journal of Bio-and Tribo-Corrosion, 2021; 7: 1–13.

[67]   Idris MB, Khalid KDU, Muhammad NA, Bala A. Corrosion inhibition and adsorption properties of Prosopis juliflora leaves extract for the corrosion of mild steel in 1M HCl solution. Journal of Scientific Research and Reports, 2016; 10: 1–7.

[68]   Zulfareen N, Kannan K, Venugopal T. Effect of methanol extract of prosopis juliflora on mild steel corrosion in 1M HCl. Chemistry & Chemical Technology, 2016; 10: 115–124.

[69]   Fouda A, El-Abbasy H, El-Sherbini A. Inhibitive effect of artemisia judaica herbs extract on the corrosion of carbon steel in hydrochloric acid solutions. International Journal of Corrosion and Scale Inhibition, 2018; 7: 213–235.

[70] Alibakhshi E, Ramezanzadeh M, Bahlakeh G, Ramezanzadeh B, Mahdavian M, Motamedi M. Glycyrrhiza glabra leaves extract as a green corrosion inhibitor for mild steel in 1 M hydrochloric acid solution: Experimental, molecular dynamics, Monte Carlo and quantum mechanics study. Journal of Molecular Liquids, 2018; 255: 185–198.

[71] Parthipan P, Cheng L, Rajasekar A. Glycyrrhiza glabra extract as an eco-friendly inhibitor for microbiologically influenced corrosion of API 5LX carbon steel in oil well produced water environments. Journal of Molecular Liquids, 2012; 333: 115952.

[72] Vasyliev G, Vorobyova V, Zhuk T. Raphanus sativus L. extract as a scale and corrosion inhibitor for mild steel in tap water. Journal of Chemistry, 2020; 2020: 1–9.

[73] Jiang X, Lai C, Xiang Z, Yang Y-f, Tan B, Long Z-q, Chen X. Study on the extract of Raphanus Sativus L. as green corrosion inhibitor for Q235 steel in HCl solution. International Journal of Electrochemical Science, 2018; 13: 3224–3234.

[74] Al-Moghrabi R, Abdel-Gaber A, Rahal H. A comparative study on the inhibitive effect of Crataegus oxyacantha and Prunus avium plant leaf extracts on the corrosion of mild steel in hydrochloric acid solution. International Journal of Industrial Chemistry, 2018; 9: 255–263.

[75] Cisneros MGV, Torres AR, Saldaña-Hereida A, Salinas-Sánchez DO Behavior of Prunus persica as green and friendly corrosion inhibitor for corrosion protection Intech Open Book Series, 2021.

[76] Raghavendra N, Bhat JI. Chemical components of mature areca nut husk extract as a potential corrosion inhibitor for mild steel and copper in both acid and alkali media. Chemical Engineering Communications, 2018; 205: 145–160.

[77] Saxena A, Prasad D, Haldhar R, Singh G, Kumar A. Use of Sida cordifolia extract as green corrosion inhibitor for mild steel in 0.5 M H_2SO_4. Journal of Environmental Chemical Engineering, 2018; 6: 694–700.

[78] Bhuvaneswari T, Vasantha V, Jeyaprabha C. Pongamia pinnata as a green corrosion inhibitor for mild steel in 1N sulfuric acid medium. Silicon, 2018; 10: 1793–1807.

[79] Akram M, Nimesh S, Chishti MA. Pongamia pinnata: An updated review on its phytochemistry, & pharmacological uses. Pharmacy & Pharmacology International Journal, 2021; 9: 194–199.

[80] Karmakar B, Samanta S, Halder G. Delonix regia heterogeneous catalyzed two-step biodiesel production from Pongamia pinnata oil using methanol and 2-propanol. Journal of Cleaner Production, 2020; 255: 120313.

[81] Vorobyova V, Skiba M, Trus I. Apricot pomaces extract (Prunus armeniaca l.) as a highly efficient sustainable corrosion inhibitor for mild steel in sodium chloride solution. International Journal of Corrosion and Scale Inhibition, 2019; 8: 1060–1083.

[82] Silva ALP, Prata JC, Walker TR, Duarte AC, Ouyang W, Barcelòand Rocha-Santos D. Increased plastic pollution due to COVID-19 pandemic: Challenges and recommendations. Chemical Engineering Journal, 2021; 405: 126683.

[83] Espejo W, Celis JE, Chiang G, Bahamonde P. Environment and COVID-19: Pollutants, impacts, dissemination, management and recommendations for facing future epidemic threats. Science of the Total Environment, 2020; 747: 141314.

[84] Anastas PT, Warner JC. Principles of green chemistry. Green Chemistry: Theory and Practice, 1998; 29: 1–27.

[85] Taha M, Foda M, Shahsavari E, Aburto-Medina A, Adetutu E, Ball A. Commercial feasibility of lignocellulose biodegradation: Possibilities and challenges. Current Opinion in Biotechnology, 2016; 38: 190–197.

[86] Stenzel F, Greve P, Lucht W, Tramberend S, Wada Y, Gerten D. Irrigation of biomass plantations may globally increase water stress more than climate change. Nature Communications, 2021; 12: 1–9.

[87] Haldhar R, Prasad D, Mandal N, Benhiba F, Bahadur I, Dagdag O. Anticorrosive properties of a green and sustainable inhibitor from leaves extract of Cannabis sativa plant: Experimental and theoretical approach. Colloids and Surfaces. A, Physicochemical and Engineering Aspects, 2021; 614: 126211.

[88] Berdimurodov E, Kholikov A, Akbarov K, Guo L, Abdullah AM, Elik M. A gossypol derivative as an efficient corrosion inhibitor for St2 steel in 1 M HCl+ 1 M KCl: An experimental and theoretical investigation. Journal of Molecular Liquids, 2021; 328: 115475.

[89] Alrefaee SH, Rhee KY, Verma C, Quraishi MA, Ebenso EE. Challenges and advantages of using plant extract as inhibitors in modern corrosion inhibition systems: Recent advancements. Journal of Molecular Liquids, 2021; 321: 114666.

[90] Akbarzadeh S, Ramezanzadeh M, Ramezanzadeh B, Bahlakeh G. Detailed atomic/molecular-level/electronic-scale computer modeling and electrochemical explorations of the adsorption and anti-corrosion effectiveness of the green nitrogen-based phytochemicals on the mild steel surface in the saline solution. Journal of Molecular Liquids, 2020; 319: 114312.

[91] Verma C, Ebenso E, Quraishi M. Alkaloids as green and environmental benign corrosion inhibitors: An overview. International Journal of Corrosion and Scale Inhibition, 2019; 8: 512–528.

[92] Sanni O, Popoola A, Fayomi O. Adsorption ability and corrosion inhibition mechanism of agricultural waste on stainless steel in chloride contaminated environment. Materials Today: Proceedings, 2021; 43: 2215–2221.

[93] Sanni O, Fayomi OSI, Popoola API. Corrosion inhibition comparison of the effect of green inhibitor on the corrosion behavior of 316L and 904L austenitic stainless steels in chloride environment. Journal of Physics. Conference Series, 2019; 1378: 022087.

[94] Hamadi L, Mansouri S, Oulmi K, Kareche A. The use of amino acids as corrosion inhibitors for metals: A review. Egyptian Journal of Petroleum, 2018; 27: 1157–1165.

[95] Saratha R, Meenakshi R. Corrosion inhibitor – A plant extract. Der Pharma Chemica, 2010; 2: 287–294.

[96] Mota MIF, Rodrigues Pinto PC, Loureiro JM, Rodrigues AE. Recovery of vanillin and syringaldehyde from lignin oxidation: A review of separation and purification processes. Separation & Purification Reviews, 2015; 45: 227–259.

[97] Yaro AS, Khadom AA, Wael RK. Garlic powder as a safe environment green corrosion inhibitor for mild steel in acidic media; adsorption and quantum chemical studies. Journal of the Chinese Chemical Society, 2014; 61: 615–623.

[98] Fadhil AA, Khadom AA, Ahmed SK, Liu H, Fu C, Mahood HB. Portulaca grandiflora as new green corrosion inhibitor for mild steel protection in hydrochloric acid: Quantitative, electrochemical, surface and spectroscopic investigations. Surfaces and Interfaces, 2020; 20: 100595.

[99] Sulaiman KO, Onawole AT, Faye O, Shuaib DT. Understanding the corrosion inhibition of mild steel by selected green compounds using chemical quantum based assessments and molecular dynamics simulations. Journal of Molecular Liquids, 2019; 279: 342–350.

[100] Parthipan P, Cheng L, Rajasekar A. Glycyrrhiza glabra extract as an eco-friendly inhibitor for microbiologically influenced corrosion of API 5LX carbon steel in oil well produced water environments. Journal of Molecular Liquids, 2021; 333: 115952.

[101] Naseri E, Hajisafari M, Kosari A, Talari M, Hosseinpour S, Davoodi A. Inhibitive effect of Clopidogrel as a green corrosion inhibitor for mild steel; statistical modeling and quantum Monte Carlo simulation studies. Journal of Molecular Liquids, 2018; 269: 193–202.

[102] Chen Z, Fadhil AA, Chen T, Khadom AA, Fu C, Fadhil NA. Green synthesis of corrosion inhibitor with biomass platform molecule: Gravimetrical, electrochemical, morphological, and theoretical investigations. Journal of Molecular Liquids, 2021; 332: 115852.

[103] Hema K, Sirajunnisa A, Venkatraman B, Subramania A. The effect of Solanam xanthocarpum leaves extract on corrosion inhibition of carbon steel in acidic medium. International Journal of Advanced Research in Chemical Sciences, 2015; 2: 10–20.

[104] Haldhar R, Prasad D, Bahadur I, Dagdag O, Kaya S, Verma DK, Kim SC. Investigation of plant waste as a renewable biomass source to develop efficient, economical and eco-friendly corrosion inhibitor. Journal of Molecular Liquids, 2021; 335: 116184.

[105] Sobhy MA, Abbas M, El-Zomrawy AA. Evaluation of aloe vera gel extract as eco-friendly corrosion inhibitor for carbon steel in 1.0 M HCl. Egyptian Journal of Chemistry, 2021; 64: 5–6.

[106] Singh AK, Mohapatra S, Pani B. Corrosion inhibition effect of Aloe Vera gel: Gravimetric and electrochemical study. Journal of Industrial and Engineering Chemistry, 2016; 33: 288–297.

[107] Sanaei Z, Ramezanzadeh M, Bahlakeh G, Ramezanzadeh B. Use of Rosa canina fruit extract as a green corrosion inhibitor for mild steel in 1 M HCl solution: A complementary experimental, molecular dynamics and quantum mechanics investigation. Journal of Industrial and Engineering Chemistry, 2019; 69: 18–31.

[108] Alibakhshi E, Ramezanzadeh M, Haddadi SA, Bahlakeh G, Ramezanzadeh B, Mahdavian M. Persian Liquorice extract as a highly efficient sustainable corrosion inhibitor for mild steel in sodium chloride solution. Journal of Cleaner Production, 2019; 210: 660–672.

[109] Li X, Deng S. Synergistic inhibition effect of walnut green husk extract and potassium iodide on the corrosion of cold rolled steel in trichloroacetic acid solution. Journal of Materials Research and Technology, 2020; 9: 15604–15620.

[110] Wu Y, Zhang Y, Jiang Y, Li N, Zhang Y, Wang L, Zhang J. Exploration of walnut green husk extract as a renewable biomass source to develop highly effective corrosion inhibitors for magnesium alloys in sodium chloride solution: Integrated experimental and theoretical studies. Colloids and Surfaces. A, Physicochemical and Engineering Aspects, 2021; 626: 126969.

[111] Ituen E, James A, Akaranta O, Sun S. Eco-friendly corrosion inhibitor from Pennisetum purpureum biomass and synergistic intensifiers for mild steel. Chinese Journal of Chemical Engineering, 2016; 24: 1442–1447.

[112] Talema A. Review of phytochemical analysis of selected traditional medicinal plants in ethiopia. International Journal of Homeopathy & Natural Medicines, 2020; 6: 23.

[113] Hamilton-Amachree A, Iroha NB. Corrosion inhibition of API 5L X80 pipeline steel in acidic environment using aqueous extract of Thevetia peruviana. Chemistry International, 2020; 6: 110–121.

[114] Ayoola AA, Obanla OR, Abatan OG, Fayomi OSI, Akande IG, Agboola O, Augustine O. Corrosion Inhibitive behaviour of the natural honey in acidic medium of A315 mild and 304 austenitic stainless steels. Analytical and Bioanalytical Electrochemistry, 2020; 12: 21–35.

[115] El-Etre AY, Abdallah M. Natural honey as corrosion inhibitor for metals and alloys. II. C-steel in high saline water. Corrosion Science, 2000; 42: 731–738.

[116] Singh A, Ahamad I, Singh VK, Quraishi MA. Inhibition effect of environmentally benign Karanj (Pongamia pinnata) seed extract on corrosion of mild steel in hydrochloric acid solution. Journal of Solid State Electrochemistry, 2011; 15: 1087–1097.

[117] Jena J, Gupta AK. Ricinus communis Linn: A phytopharmacological review. International Journal of Pharmacy and Pharmaceutical Sciences, 2012; 4: 25–29.

[118] Minhaj A, Saini PA, Quraishi MA, Farooqi IH. A study of natural compounds as corrosion inhibitors for industrial cooling systems. Bottorff of Titntt, 1999; 373: 32–36.

[119] Pal A, Das C. A novel use of solid waste extract from tea factory as corrosion inhibitor in acidic media on boiler quality steel. Industrial Crops and Products, 2020; 151: 112468.

[120] Hasan AM, Abdel-Raouf ME. Applications of guar gum and its derivatives in petroleum industry: A review. Egyptian Journal of Petroleum, 2018; 27: 1043–1050.

[121] Belakhdar A, Ferkous H, Djellali S, Sahraoui R, Lahbib H, Amor YB, Benguerba Y. Computational and experimental studies on the efficiency of Rosmarinus officinalis polyphenols as green corrosion inhibitors for XC48 steel in acidic medium. Colloids and Surfaces. A, Physicochemical and Engineering Aspects, 2020; 606: 125458.

[122] Prabha AS, Kavitha K, Rajendran S. Inhibition of corrosion of mild steel in simulated oil well water by an aqueous extract of Andrographis paniculata. Indian Journal of Chemical Technology, 2021; 27: 452–460.

[123] Boukroufa M, Boutekedjiret C, Petigny L, Rakotomanomana N, Chemat F. Bio-refinery of orange peels waste: A new concept based on integrated green and solvent free extraction processes using ultrasound and microwave techniques to obtain essential oil, polyphenols and pectin. Ultrasonics Sonochemistry, 2015; 24: 72–79.

[124] de Oliveira GA, Teixeira VM, da Cunha JN, Rangel M, dos Santos VMP, Rezende MJC, D'Elia E. Biomass of microalgae chlorella sorokiniana as green corrosion inhibitor for mild steel in HCl solution. International Journal of Electrochemical Science, 2021; 16: 2021.

[125] Rodrigues LS, do Valle AF, D'Elia E. Biomass of microalgae Spirulina maxima as a corrosion inhibitor for 1020 carbon steel in acidic solution. International Journal of Electrochemical Science, 2013; 13: 6169–6189.

[126] Farhadian A, Rahimi A, Safaei N, Shaabani A, Sadeh E, Abdouss M, Alavi A. Exploration of sunflower oil as a renewable biomass source to develop scalable and highly effective corrosion inhibitors in a 15% HCl medium at high temperatures. ACS Applied Materials & Interfaces, 2021; 13: 3119–3138.

[127] Singh A, Ebenso EE, Qurashi MA. Corrosion inhibition of carbon steel in HCl solution by some plant extracts. International Journal of Corrosion, 2012; 2012: 1–20.

[128] Karim ARM, Susanto R, Azhari I, Utami SP, Heltina D. Biomass extraction as green corrosion inhibitor for low carbon steel in hydrochloric acid solution. Journal of Physics. Conference Series, 2020; 1655: 012038.

[129] Odewunmi NA, Umoren SA, Gasem ZM. Utilization of watermelon rind extract as a green corrosion inhibitor for mild steel in acidic media. Journal of Industrial and Engineering Chemistry, 2015; 21: 239–247.

[130] Rahayu PP, Sundari CDD, Farida I. Corrosion inhibition using lignin of sugarcane bagasse. IOP Conference Series: Materials Science and Engineering, 2018; 434: 012087.

[131] Salleh SZ, Yusoff H, Zakaria SK, Taib MAA, Abu Seman A, Masri MN, Mohamad M, Mamat S, Sobri SA, Ali A, Teo PT. Plant extracts as green corrosion inhibitor for ferrous metal alloys: A review. Journal of Cleaner Production, 2021; 304: 127030.

[132] Reza NA, Akhmal NH, Fadil NA, Taib MFM. A review on plants and biomass wastes as organic green corrosion inhibitors for mild steel in acidic environment. Metals, 2021; 11: 1062.

[133] Marzorati S, Verotta L, Trasatti SP. Green corrosion inhibitors from natural sources and biomass wastes. Molecules, 2019; 24: 48.

Younes Ahmadi, Mubasher Furmuly, Nasrin Raji Popalzai

6 Biopolymers as corrosion inhibitors for metals in corrosive media

Abstract: Biopolymers or bio-based polymers are biodegradable macromolecules that are found in abundance, cost-effective, nontoxic, and are eco-friendly. They are considered to be one of the most promising substitutes to compete for petro-based materials. The outstanding features of biopolymers, such as exceptional chemical architecture and a large number of active sites (e.g., hydroxyls, esters, carboxyl, and other functionalities), make them suitable candidates for anticorrosive materials. In addition, such characteristics enabled their facile chemical functionalization and composite formation to produce effective anticorrosive materials, which can withstand highly corrosive environments. Therefore, this chapter has been designed to summarize the applicability of biopolymers, such as chitosan, carboxyl methylcellulose, gum arabic, xanthan gum, and alginates, as anticorrosive materials. Further, the anticorrosive action mechanism of biopolymers and their composites in an aggressive medium has been discussed in this chapter.

Keywords: Anticorrosion, biopolymers, eco-friendly: sustainable, chitosan, coatings

6.1 Introduction

Corrosion is the deterioration of materials (e.g., metals) that leads to a loss of their properties. Corrosion generally occurs due to the reaction of materials with their surrounding environment [1]. According to the corrosion societies, corrosion and its consequences cost around 5% of the GDP of developed nations [2]. Several protecting methods (such as cathodic protection, coating of materials, and application of corrosion inhibitors) are generally employed to combat or control corrosion [3]. Among the various protecting approaches, the utilization of protective coatings has attracted immense attention. The coatings generally form a thin layer over the metal substrates that prevent their reaction with their surrounding environment, thereby protecting them from corrosion and enhancing their lifetime [4]. Ideal anticorrosive coatings provide effective physical barriers that impede the aggressive species (and ions) from reaching the metal interface [5]. Conventionally, effective anticorrosion systems required the application of chromate-rich surface treatments (primer- and pigment-based chromates) and the application of petro-based polymers [6, 7]. However, environmental concern has forced scientists and technologists to develop green and eco-friendly materials and methods for the effective protection of metals. As a result, several alternative approaches have been explored, which

https://doi.org/10.1515/9783110760583-006

include a wide range of "green" surface treatments and pretreatments, the use of environmental-friendly pigments, and natural anticorrosive materials [8]. The latest progress is the production and application of eco-friendly coatings that produce/ use a low amount of volatile organic compounds (VOCs). Generally, such coatings have been developed via waterborne formulations, isocyanate-free nanocomposites, smart and self-healing polymers, and the application of bio-based polymers [9, 10]. Among these methods, the application of biodegradable, biocompatible, and environmentally friendly polymeric or composite coatings is the most promising alternative to combat the corrosion of metals [7]. Utilization of biopolymers improved the mechanical, chemical, adhesion, barrier, thermal, and anticorrosive properties of the resultant polymer coatings. In addition, the presence of a large number of functional groups in the structure of biopolymers offered a suitable environment for the dispersion of nanofillers [11].

Biopolymers are a class of naturally derived bio-macromolecules that can be classified into three groups – polysaccharides, polypeptides/protein, and polynucleotides [12]. Among these, polysaccharides (e.g., cellulose, chitosan, and alginate) and their derivatives have played a significant role in the development of anticorrosive coatings [8, 13, 14]. These biopolymers are abundantly available and are generally derived from natural and renewable sources, such as plants, plant seeds, and marine organisms [15]. These polymers are flexible and biodegradable, possessing a large number of reactive sites; hence they have attracted huge attention in various application realms [16]. The applicability of natural and synthetic polymers as anticorrosive coatings (for metals) has been explored, confirming their potency in the protection of metals against corrosion [16]. However, the pure form of these polymers exhibited moderate corrosion inhibition effects [17]. Therefore, several attempts, including the incorporation of substances with synergistic effects, copolymerizing, blending, crosslinking, and addition of (nano) fillers into the polymer matrix, have been made [18–20]. These approaches have improved the inhibition ability of the resultant coating materials. The utilization of anticorrosive composites and nanocomposites has shown superior protecting behavior, which barricaded the metal surfaces from corrosive species [21]. In the light of the above discussion, the present chapter aims to highlight the application of biopolymers and their (nano) composites as anticorrosive materials for the protection of metals.

6.2 Anticorrosive biopolymer-based composite and nanocomposite coatings

The quest for developing anticorrosive materials for the protection of metals has required the application of eco-friendly natural biopolymer-based composites and nanocomposite coatings. These corrosion inhibitor coatings have been increasingly used and examined for various substrates under aggressive corrosive media. Compositing is generally referred to as a modification technique, which includes the reaction(s) of different precursors to formulate a new material, with chemically distinct characteristics different from the pristine components. However, in the case of nanocomposite systems, the dimension of one of the components falls in the range of nanometers (generally < 0.1 µm) [11].

6.2.1 Chitosan-based anticorrosive coatings

Chitosan is abundant in nature and is present in the exoskeletons of marine organisms, such as shrimp and crabs [22]. Chitosan can also be obtained from simple arthropods and the N-deacetylation of chitin present in the cell wall of fungi through alkaline treatment [23]. The chitosan biopolymer possesses (1,4)-β-N-acetyl glycosaminoglycan as the repeating monomer (Figure 6.1) [23, 24]. In addition, chitosan has a good antiseptic property against bacteria and fungi, which facilitates its application in coating formulation [23]. Chitosan and its derivatives are used as effective inhibitors for the protection of various metals and alloys under harsh environments. The anticorrosive performance of chitosan-based coatings strongly depends on the substrate and the nature of the corrosive environment. For instance, under acidic conditions (e.g., 0.5 M HCl), chitosan-based coatings could effectively protect copper metal against corrosion (inhibition efficiency = 93%) [25]. However, the biopolymer exhibited poor performance for the protection of steel under similar conditions.

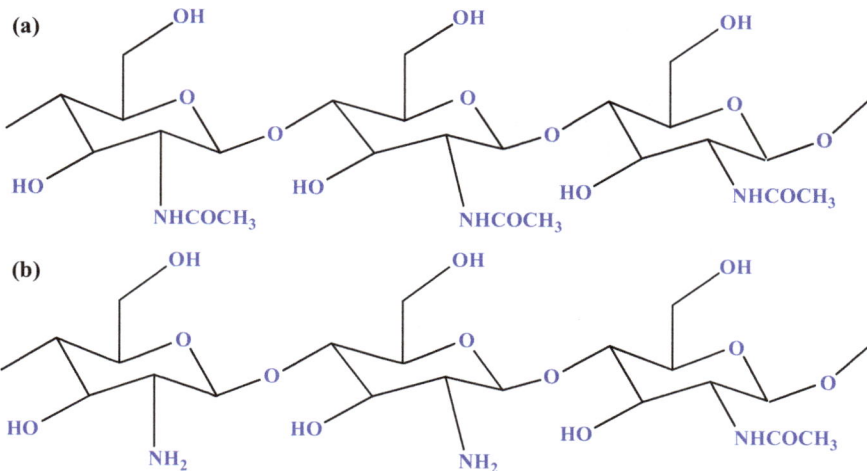

Figure 6.1: Structure of (a) chitin and (b) chitosan (adapted with permission from reference [24], Copyright (2020), Elsevier).

The formulation of anticorrosive biopolymer-based composites for the protection of metals improved the desired properties of the resultant coatings. These properties include chemical stability, mechanical strength, robust adhesion, and superior barrier performance, which are vital for effective inhibitory biopolymer coatings. Keeping this in mind, composite and nanocomposite-based anticorrosive chitosan coatings were successfully developed using various classes of nanofillers [24, 25]. Fillers were incorporated in the chitosan matrix, ex situ, in situ, sol–gel, and by electrochemical deposition methods [18, 26, 27]. In the case of the *ex-situ* approach, the preparation of fillers occurs outside the corrosive medium, before their dispersal into the biopolymers matrix. While in the case of the *in situ* technique (as the name suggests), nanofillers are dispersed during the formulation of the biopolymer coatings [28]. The in situ approach offers various advantages, such as homogeneous dispersion of fillers that prevent their agglomeration within the matrix. In the case of the sol–gel methodology, solid nanoparticles are dispersed within the monomer solution, forming a colloidal suspension (i.e., sol). Subsequently, the interconnecting network among the gel and the sol phases is formed through a polymerization reaction, followed by hydrolysis (Figure 6.2) [27].

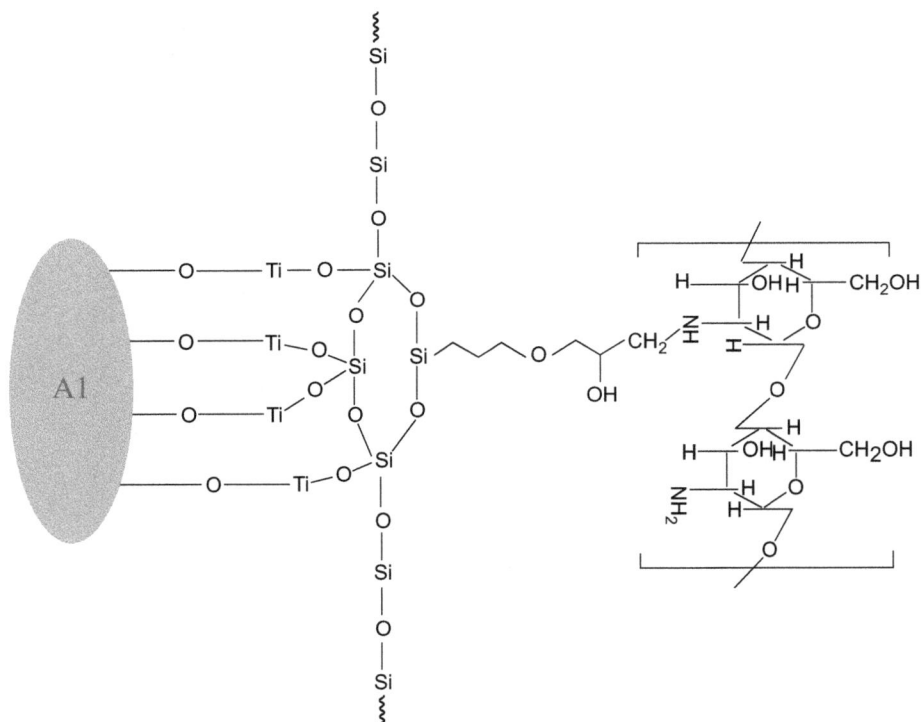

Figure 6.2: Scheme representing chitosan-based TiO_2 nanocomposite coated on the Al surface (adapted with permission from reference [27], Copyright (2017), Elsevier).

The electrochemical deposition technique is generally used for the formulation of anticorrosive composite/nanocomposite coatings. In this case, the protective films are deposited on the surface of substrates to protect them from corrosive environments. The anticorrosive composites are applied to the metal's surface as inhibitors or coatings. For instance, the poly (N-vinyl imidazole) grafted carboxymethyl chitosan composites (CMCh-g-PVI) have been developed for the protection of steel (API X70) in acidic conditions [26]. The formulated composites significantly prevented the corrosion of the metal substrate, compared to the pristine chitosan and carboxymethyl chitosan under a similar environment. The chitosan-based anticorrosive coatings, embedded with different corrosion inhibitors (like mercaptobenzothiazole (MBT) and benzotriazole (BTA)), were also prepared for the protection of copper-based alloys [29]. The efficacy of the formulated coatings was examined by experimenting with their corrosion protection behavior under a corrosive medium. The coated and bare alloys were characterized after the corrosion studies. They revealed corrosion resistance properties of the formulated MBT and BTA containing chitosan-based coatings (Figure 6.3).

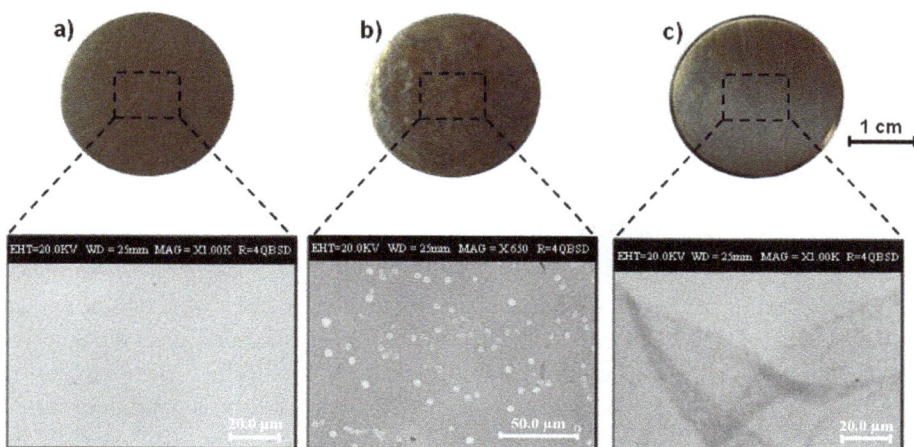

Figure 6.3: The SEM images and photographs of (a) bare metal and coated metal with chitosan film prepared from (b) acetic acid and (c) D-(+)-gluconic δ-lactone solutions (adapted with permission from reference [29], Copyright (2018), Elsevier).

Polymeric composite/nanocomposites are deployed as coatings and not corrosion inhibitors in most of the corrosion mitigation practices [30, 31]. This is mainly due to the fact that polymers are generally insoluble in aqueous solutions, which plays a vital role in the corrosion inhibition property of inhibitors. A completely soluble inhibitor can perform better than partially soluble inhibitors [32]. In addition, the application of appropriate fillers plays a significant role in the protection property of chitosan-based anticorrosive coatings. Recently, the potential of 1-carboxymethyl-3-

methylimidazolium bis(trifuoromethylsulfonyl) imide, and benzotriazole additives was explored in the preparation of chitosan-based 1-carboxymethyl-3-methylimidazolium bis(trifluoromethylsulfonyl) imide, that is, $HO_2CC_1MImNTf_2$-chitosan) anticorrosive coatings [33]. The anticorrosion properties of the fabricated materials were tested under corrosive environments (HCl vapor), which revealed that the $HO_2CC_1MImNTf_2$-chitosan coating exhibited lesser effectiveness than the benzotriazole-chitosan coatings. However, these coatings maintained high transparency without affecting the appearance of the substrate (bronze) (Figure 6.4).

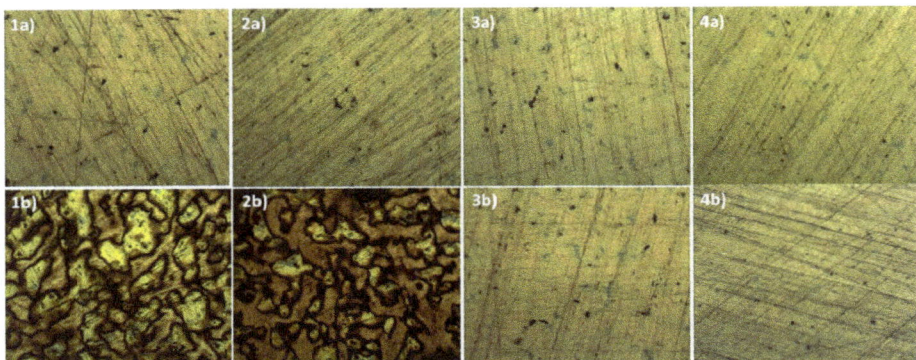

Figure 6.4: The optical micrographs of bronze metal coated with chitosan (1), chitosan/ $HO_2CC_1MImNTf_2$ in the water (2), chitosan-benzotriazole (3), and chitosan/$HO_2CC_1MImNTf_2$/ benzotriazole (4). The (a) and (b) series are for before and after exposure to HCl vapor (100× magnification) (adapted with permission from reference [33], Copyright (2020), Springer).

Chitosan-based Schiff bases (C-SB) have also been used for the prevention of metals against corrosion. For instance, C-SB was used to protect mild steel against the corrosive environment (1 M HCl solution) [34]. The protection efficiency was determined by examining the weight loss, electrochemical techniques (such as potentiodynamic polarization (PDP) and electrochemical impedance spectroscopy (EIS)), energy dispersive X-ray (EDX), and microscopy techniques (scanning electron microscopy (SEM), atomic force spectroscope (AFM)). The study was conducted at different temperatures and concentrations, which suggested the inhibition performance of C-SB up to 86.94% and 333 K temperature. Electrochemical studies suggested that C-SB acts as an interface as well as a mixed-type corrosion inhibitor that limited the cathodic and anodic reactions. Later, chemically modified chitosan-formaldehyde SBs using thiocarbohydrazide (TC) and thiosemicarbazide (TS) were reported [35]. The modified chitosan thiosemicarbazide (TSFCS) and thiocarbohydrazide (TCFCS) were used as anticorrosive materials for steel (under 2% solution of acetic acid) protection. Recently, chitosan nanocomposite biopolymer coatings on copper were also formulated and their anticorrosion efficiency was examined [9]. The incorporation of silica nanoparticles considerably affected the properties of chitosan by reducing the swelling ratio and enhancing the thermal stability of the formulated coatings. It has also been experimented that the corrosion resistance of chitosan nanocomposite-based coatings was enhanced by loading silica and 2-mercaptobenzothiazole within the chitosan matrix. It was revealed that the crosslinking in chitosan coatings was enhanced by the addition of nanofillers, which subsequently increased the corrosion resistance of the materials (inhibition efficiency = 85%).

6.2.2 Cellulose-based anticorrosive agents

Carboxymethyl cellulose (CMC) is an abundant and natural biopolymer, which is derived through alkali-catalyzed reaction of cellulose with chloroacetic acid [36]. CMC has been widely used as a viscosity modifier, stabilizer, and thickener in the food industry, for example, ice creams and other food products [37, 38]. CMC exists as a sodium salt that resembles cellulose in terms of structural peculiarities, while differing in the alignment of the reactive carboxymethyl groups. In the case of sodium salt of CMC, the carboxymethyl functionalities are bound to the –OH groups of its cellulosic glucopyranyl moieties (Figure 6.5) [39]. CMC is also an important ingredient in other products, such as detergents, textile sizing, diet pills, paper, laxatives, toothpaste, reusable heat packs, and water-based paints [39].

Figure 6.5: Synthesis and chemical structure of CMC (adapted with permission from reference [39], Copyright (2019), Elsevier).

In the coating industry, CMC has been applied and examined as a corrosion inhibition material in acid as well as saline environments [40, 41]. However, it was found that CMC possesses a moderate inhibiting effect against corrosive environments. Nevertheless, the incorporation of nanoparticles within the CMC matrices could boost the protection efficiency of the resultant CMC nanocomposite coatings. For instance, it was observed that CMC was able to moderately protect (64.8%) the mild steel under the acidic condition (2 M H_2SO_4) at 303 K [40]. In addition, the protection efficiency of CMC was reduced at higher temperatures [40]. Further, it was found that the incorporation of nanoparticles (silver nanoparticles) in the CMC matrix enhanced the corrosion protection of the material under an acidic environment

(15% H_2SO_4 solution) by 86.4% at 313 K [42]. Interestingly, the study revealed that the inhibition efficiency of the obtained nanocomposite coatings increased by raising the temperature. These behaviors of pristine and nanocomposite-based CMC anticorrosive materials can be explained by the extent of interactions between the anions present in the corrosive medium and in the CMC-based inhibitors. For instance, the metal surfaces receive positive charges and undergo hydration with anions under an acidic medium (e.g., sulfate anions in H_2SO_4 solution with concentration ≥1). As a result, pristine CMC existed as poly-cations under strongly acidic conditions, meaning that the as-formed CMC poly-cations were adsorbed on the metal interfaces by interacting with the sulfate anions, proposing the physisorption mechanism. The sulfate anions possess smaller shielding power (compared to the chloride anions), which does not allow them to fully replenish the surface, making it harder for CMC poly-cations to be adsorbed; hence moderate corrosion inhibition was observed. It should be noted that the rod-like structure of CMC under high pH solutions and temperature overlap (agglomerate), coil up, entangle, and form thermoreversible gels, which may have been the reason for lower inhibition efficiency under 2 M H_2SO_4 solution when temperature is increased. Instead, the active properties of nanoparticles facilitate the formation of chemical interactions between nanocomposites and the metal surfaces that results in the suppressing of the opposing forces of positively charged metal surfaces. As a result, a larger number of polymers participate in the adsorption process, which makes the nanocomposites better inhibitors than the pristine biopolymers. In addition, the presence of nanoparticles within the CMC network could maintain the rod-like, extended, and nonentangled form even at higher temperatures, which blocked the secondary exposure of metal surfaces to corrosive attacks.

Hydroxypropyl methylcellulose acetate succinate (HPMCAS) is a derivative of hydroxypropyl methylcellulose biopolymer [13]. HPMCAS is biodegradable and eco-friendly material, which has also been used for the corrosion protection of metals [13]. In this regard, the electrochemical properties of HPMCAS-based anticorrosive coatings (under acidic environments) were examined using EIS and PDP studies. This study revealed that the thickness of the fabricated coatings increased due to the swelling effect, which resulted in increasing the corrosion resistance of the HPMCAS-based coatings.

Recently, self-healing anticorrosive bio-based coatings have been significantly explored and employed for the protection of metals. Such anticorrosive coatings not only improve the corrosion resistance properties but also hinder the diffusion and passage of the destructive ions through the coating film. The extrinsic smart self-healing protecting coatings are generally mixed with healing agents/inhibitors (in nano-sized containers) [43]. When the anticorrosive coatings are scratched as a result of an external stimulus, the healing agents present within the coatings are released to repair the defects, inhibiting further corrosion. Consequently, the service life of the protective coatings is extended and the anticorrosive effects are also improved. The self-healing anticorrosive coatings protect the underlying metals via various methods, such as cathodic disbonding resistance and physical protection [44]. For instance, recently, methylcellulose (MC)-based core–shell fiber self-healable anticorrosive coatings have been fabricated [45]. In this study, electro-spun MC fibers (i.e., shell) were combined with a silicone coating material to improve the deterioration protection of the underlying metal (i.e., carbon steels) (Figure 6.6). However, the self-healing property of such anticorrosive coatings should be investigated further since the formulated coatings exhibited decreased healing ability (due to the exhaustion of healing agents) upon a longer immersion period.

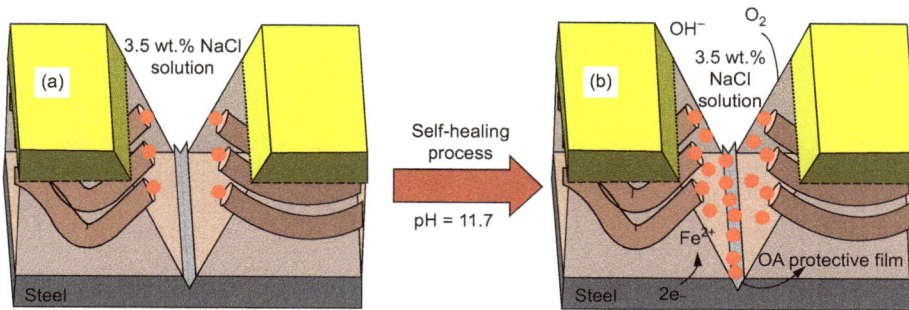

Figure 6.6: Self-healing mechanism of MC core–shell fiber anticorrosive coatings (adapted with permission from reference [45], Copyright (2021), Elsevier).

6.2.3 Other biopolymer-based anticorrosive materials

The green, eco-friendly, and cost-effectiveness of biopolymers have encouraged their application in the production of anticorrosive agents. In this regard, several composite/nanocomposites of other biopolymers have been reported, which are employed as corrosion inhibitors of metals. These biopolymer-based systems include xanthan gum (XG)/poly(acrylamide), XG/polyaniline, pectin biopolymer, alginate, and gum arabic (GA)/silver composites [10, 28, 46–48]. GA is a mixture of polysaccharides and glycoproteins, which is obtained from *Acacia* trees. The structure of GA possesses amino acids linked to short arabinose side chains [49]. The high-water solubility of GA facilitated its application as a binder in the production of water-based paints [50]. GA plays a hydrophilic and pore-forming role in the production of polysulfone membranes [51]. In addition, GA has been used for the protection of metals (Al and Steel) as corrosion inhibitors in acidic media [52, 53]. In addition, GA-based anticorrosive nanocomposites (using silver nanoparticles) were used as eco-friendly anticorrosive materials for steel protection (under 15% HCl and H_2SO_4 media) [53]. The formulated nanocomposites effectively mitigated the corrosion of the steel surface, particularly in HCl medium (Figure 6.7) [53, 54]. It was found that the GA-based nanocomposite particles protected the metal surface against corrosion by physically adsorbing on the steel surface. The application of sodium alginate for the formulation of eco-friendly waterborne polyurethane-ZnO nanocomposite anticorrosive coatings has been reported [28]. The coatings were formulated through the ultra-sonication method and were effectively used for protecting the mild steel. The corrosion resistance of coated steel was investigated by PDP and EIS studies, in which a small amount of lignosulfonate-modified ZnO (0.3 wt%) exhibited better protection than the pristine sodium alginate.

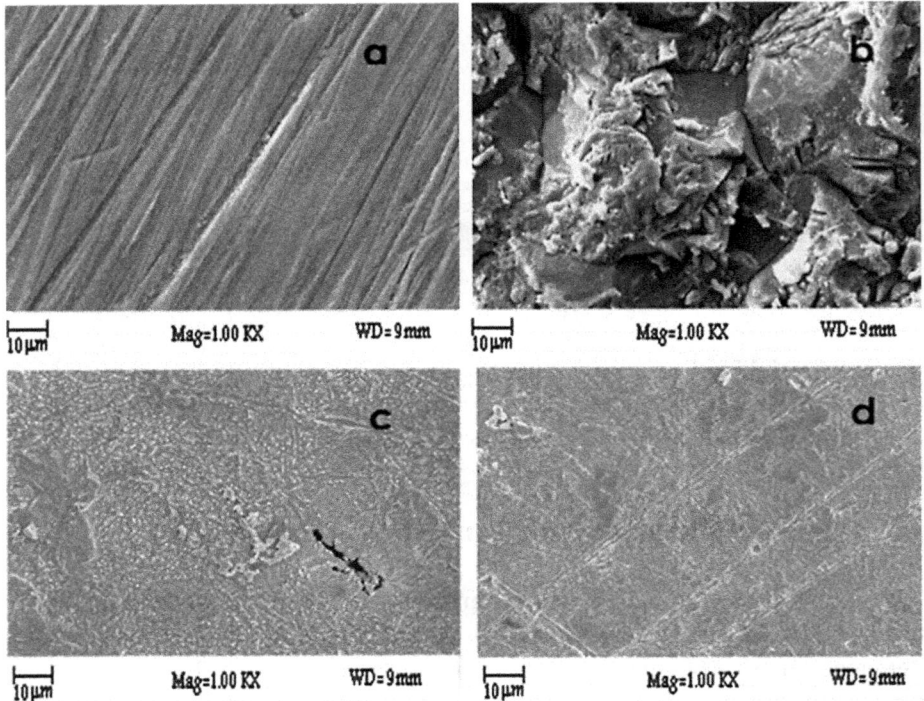

Figure 6.7: The SEM micrographs of (a) bare steel before corrosion, while (b), (c), and (d) show the steel samples upon immersion 8 h in 2 M HCl, 2 M HCl + 0.8 g/l *Ferula assa-foetida*, and 2 M HCl + 0.8 g/l *Dorema ammoniacum*, respectively (adapted with permission from reference [54], Copyright (2011), Elsevier).

XG (i.e., a polysaccharide) consists of a main chain of β-(1/4)-D-glycopyranose residues with side chains of trisaccharide possessing (4/1)-β-D-mannopyranose, (2/1)-β-D-glucuronic acid, and (3/1)-β-D-mannopyranose [55]. XG is also extensively used as a thickening and stabilizing agent in food products. It has also been used as a corrosion inhibitor for the protection of metals [53]. For instance, the grafting of the XG backbone with polyaniline (PANI) resulted in the formation of conducting biopolymer composites that could protect the metal surface against corrosion under acidic conditions (1 M HCl solution) [56]. The impedance measurements revealed that the corrosion protection efficiency of XG was 65.89% while that of XG-g-PANI was found to be 95.32% [57].

6.3 Anticorrosive mechanism of biopolymer-based materials

The modified biopolymers (as composites or nanocomposites) maintain their π-electrons and heteroatoms, which act as adsorption centers [42]. Generally, organic molecules physically and/or chemically adsorb on metal surfaces [58]. In the case of physical adsorption, the charged inhibitor molecules are attracted to the charged surface of metals through electrostatic forces, while in the case of chemical adsorption, the electron pairs of inhibitor molecules are transferred/shared with the empty d-orbitals of the metal substrate. Therefore, several factors, including pH, the form of molecules, anions, temperature, and the charge of the metal surface, play critical roles in the mechanism of inhibition. The charge of the substrate surface is generally determined by measuring the difference between the E_{corr} (corrosion potential) and the potential of zero charges ($E_{q=0}$). The positively and negatively charged metal surfaces have $E_{corr} - E_{q=0} = -d$ and $E_{corr} - E_{q=0} = +d$, respectively [59]. The positively charged metal surface would be covered with the anion molecules present in the corrosive environment. However, if the inhibitor particles are in their protonated state, physical adsorption will be the dominant mechanism. Inhibitors having a neutral state in the corrosive medium generally undergo a chemical adsorption mechanism. It should be noted that both mechanisms can occur simultaneously on the metal surface [60].

Recently, cellulose–lignin-based barrier composite coatings, formulated through aqueous electrophoretic deposition (at 0.5 V), were used for the protection of hot-dip galvanized (HDG) steel [61]. The coalescence of colloidal lignin particles occurred during the drying process of the coatings, which resulted in the fabrication of a protecting film on the surface of the HDG-steel. It was further observed that the compact and impermeable nature of the protecting layers diminished the diffusion of electrolytes and corrosive species to the metal-coating boundary for about 48 h of exposure to the corrosive environment (3.5% NaCl) (Figure 6.8). In addition, EIS investigation revealed that the formulated protective composite coatings delivered a high charge transfer resistance, compared to bare steel, even after immersion for 15 days (13.7 cf. 0.2 $k\Omega \cdot cm^2$). Moreover, the utilization of the electrophoretic deposition approach appeared to be an effective approach for the formulation of adhesive bio-based coatings.

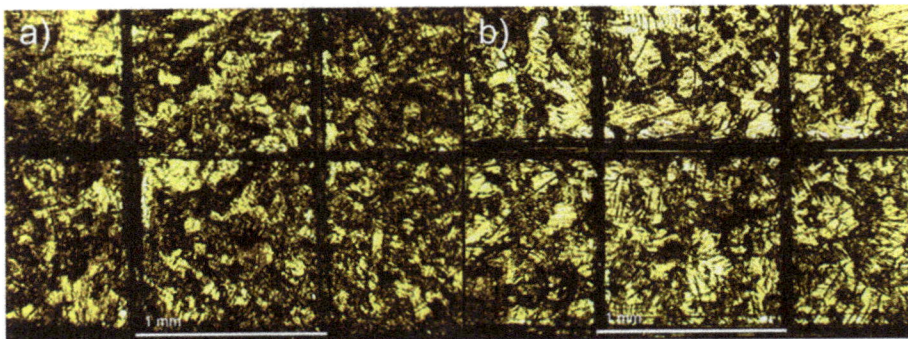

Figure 6.8: Post cross-cut measurements: Optical microscopy images of composite coating surfaces containing (a) 1 and (b) 2 g/L oxidized cellulose nanofibers suspension deposited at 0.5 V (scale bar: 1 mm) (adapted with permission from reference [61], Copyright (2021), American Chemical Society).

6.4 Conclusion

The chapter summarized the potential utilization of various biopolymers along with their composites and nanocomposites, such as chitosan, carboxyl methylcellulose, GA, XG, and alginates, as anticorrosive materials under different corrosive media. Based on this discussion, it can be concluded that the anticorrosive coatings formulated from biopolymers and their derivatives are ideal environmentally benign candidates that can effectively substitute the conventional toxic corrosion inhibitors. The presence of a large number of functional groups in biopolymers provide them with outstanding surface protection and coverage ability, which facilitate their good corrosion inhibiting properties for a wide class of metals and alloys. Nevertheless, the low solubility (or nonsolubility) of some biopolymers (like chitosan in the aqueous electrolytes) has limited their use in anticorrosive applications. Therefore, several effective attempts (like chemical modifications) have been made to increase their solubility. Such functionalization approaches enhanced the solubility of biopolymers and increased the protection efficiency of derivatives of such materials. In addition, biopolymers are capable of forming chelating complexes with metal ions, which helps them in providing excellent long-term anticorrosive behavior. Most of the biopolymer-based anticorrosive compounds (e.g., chitosan-based) can act as mixed-type anticorrosion materials with a small cathodic inhibition property.

Certainly, the fabrication and designing of sustainable self-healing coatings will become a research hot spot in the near future. Sustainable healing specifies that infinite repair cycles are carried out without weakening the strength of each healing cycle. These advancements may endow the self-healing mechanism to repair a larger range of cracks with the ability to regenerate the coatings, similar to that of the lizard tail. Besides, it is essential to fabricate and design durable healing agents and polymers to attain superior self-repairing and anticorrosion results by consuming a lesser amount of such agents. In order to achieve these goals, microcapsules and microvascular structures can be incorporated within the biopolymer matrix, with self-healing ability. In these cases, microcapsules can help repair small cracks, while microvascular structures and self-healable polymer coatings can repair larger wounds. Incorporation of the advantages of these materials can help achieve the improved healing effect as well as advanced and smart anticorrosion performances. Though biopolymer-based anticorrosive materials act as excellent anticorrosive agents, scanty literature on their application is available. Therefore, it is highly required to explore their consumption as environmentally friendly anticorrosive materials.

References

[1] Ahmadi Y, Yadav M, Ahmad S. Oleo-polyurethane-carbon nanocomposites: Effects of in-situ polymerization and sustainable precursor on structure, mechanical, thermal, and antimicrobial surface-activity. Composites Part B [Internet] Elsevier, 2019; 164: 683–692. Available from https://doi.org/10.1016/j.compositesb.2019.01.078

[2] Hou BX, Li X, Ma X, Du C, Zhang D, Zheng M, Xu W, Lu D, Ma F. The cost of corrosion in China. Njp Materials Degradation, Nature, 2017; 1: 4.

[3] Ahmadi Y, Ahmad S. Surface-active antimicrobial and anticorrosive Oleo-Polyurethane/graphene oxide nanocomposite coatings: Synergistic effects of in-situ polymerization and π-π interaction. Prog Org Coat [Internet] Elsevier, 2019; 127: 168–180. Available from https://doi.org/10.1016/j.porgcoat.2018.11.019

[4] Ahmadi Y, Siddiqui MT, Haq QMR, Ahmad S. Synthesis and characterization of surface-active antimicrobial hyperbranched polyurethane coatings based on oleo-ethers of boric acid. Arabian Journal of Chemistry [Internet] King Saud University, 2020; 13: 2689–2701. Available from https://doi.org/10.1016/j.arabjc.2018.07.001

[5] Ahmadi Y, Ahmad S. Formulation of a promising antimicrobial and anticorrosive bi-functional boronated hyperbranched oleo-polyurethane composite coating through the exploitation of functionalized reduced graphene oxide as chain extender. Applied Surface Science, 2019; 494: 196–210.

[6] Kopeć M, Rossenaar BD, van Leerdam K, Davies AN, Lyon SB, Visser P, et al. Chromate ion transport in epoxy films: Influence of BaSO4 particles. Progress in Organic Coatings, 2020; 147: 105739.

[7] Gharbi O, Thomas S, Smith C, Birbilis N. Chromate replacement: What does the future hold? npj Mater Degrad [Internet]. Springer US, 2018; 2: 23–25. Available from http://dx.doi.org/10.1038/s41529-018-0034-5

[8] Verma C, Quraishi MA, Alfantazi A, Rhee KY. Corrosion inhibition potential of chitosan based Schiff bases: Design, performance and applications. International Journal of Biological Macromolecules [Internet]. Elsevier B.V., 2021; 184: 135–143. Available from https://doi.org/10.1016/j.ijbiomac.2021.06.049

[9] Bahari HS, Ye F, Carrillo EAT, Leliopoulos C, Savaloni H, Dutta J. Chitosan nanocomposite coatings with enhanced corrosion inhibition effects for copper. International Journal of

Biological Macromolecules [Internet]. The Authors, 2020; 162: 1566–1577. Available from https://doi.org/10.1016/j.ijbiomac.2020.08.035

[10]   Sushmitha Y, Rao P. Material conservation and surface coating enhancement with starch-pectin biopolymer blend: A way towards green. Surfaces and Interfaces [Internet]. Elsevier, 2019; 16: 67–75. Available from https://doi.org/10.1016/j.surfin.2019.04.011

[11]   Umoren SA, Eduok UM. Application of carbohydrate polymers as corrosion inhibitors for metal substrates in different media: A review. Carbohydrate Polymers [Internet]. Elsevier Ltd., 2016; 140: 314–341. Available from http://dx.doi.org/10.1016/j.carbpol.2015.12.038

[12]   Xiong R, Grant AM, Ma R, Zhang S, Tsukruk VV. Naturally-derived biopolymer nanocomposites: Interfacial design, properties and emerging applications. Mater Sci Eng R Reports [Internet]. Elsevier B.V., 2018; 125: 1–41. Available from http://dx.doi.org/10.1016/j.mser.2018.01.002

[13]   Shi SC, Su CC. Electrochemical behavior of hydroxypropyl methylcellulose acetate succinate as novel biopolymeric anticorrosion coating. Materials Chemistry and Physics [Internet]. Elsevier B.V., 2020; 248: 122929. Available from https://doi.org/10.1016/j.matchemphys.2020.122929

[14]   Pais M, Rao P. Biomolecules for Corrosion Mitigation of Zinc: A Short Review. Journal of Bio- and Tribo-Corrosion [Internet]. Springer International Publishing, 2019; 5: 1–11. Available from https://doi.org/10.1007/s40735-019-0286-9

[15]   Shahini MH, Ramezanzadeh B, Mohammadloo HE. Recent advances in biopolymers/carbohydrate polymers as effective corrosion inhibitive macro-molecules: A review study from experimental and theoretical views. Journal of Molecular Liquids [Internet]. Elsevier B.V., 2021; 325: 115110. Available from https://doi.org/10.1016/j.molliq.2020.115110

[16]   Mekonnen TH, Haile T, Ly M. Hydrophobic functionalization of cellulose nanocrystals for enhanced corrosion resistance of polyurethane nanocomposite coatings. Applied Surface Science [Internet]. Elsevier B.V., 2021; 540: 148299. Available from https://doi.org/10.1016/j.apsusc.2020.148299

[17]   Umoren SA, Solomon MM. Synergistic corrosion inhibition effect of metal cations and mixtures of organic compounds: A Review. J Environ Chem Eng [Internet]. Elsevier B.V., 2017; 5: 246–273. Available from http://dx.doi.org/10.1016/j.jece.2016.12.001

[18]   Solomon MM, Gerengi H, Kaya T, Umoren SA. Performance Evaluation of a Chitosan/Silver Nanoparticles Composite on St37 Steel Corrosion in a 15% HCl Solution. ACS Sustainable Chemistry & Engineering, 2017; 5: 809–820.

[19]   Hefni HHH, Azzam EM, Badr EA, Hussein M, Tawfik SM. Synthesis, characterization and anticorrosion potentials of chitosan-g-PEG assembled on silver nanoparticles. International Journal of Biological Macromolecules [Internet]. Elsevier B.V., 2016; 83: 297–305. Available from http://dx.doi.org/10.1016/j.ijbiomac.2015.11.073

[20]   Solomon MM, Gerengi H, Umoren SA. Carboxymethyl Cellulose/Silver Nanoparticles Composite: Synthesis, Characterization and Application as a Benign Corrosion Inhibitor for St37 Steel in 15% H2SO4 Medium. ACS Applied Materials & Interfaces, 2017; 9: 6376–6389.

[21]   Safari R, Ehsani A, Kashi AH, Hadi H, Beladi FS. External electric field effects on electronic properties of a candidate eco-friendly biopolymer and its anticorrosive properties in acidic media. Journal of Materials Engineering and Performance [Internet]. Springer US, 2021; 30: 522–534. Available from https://doi.org/10.1007/s11665-020-05328-1

[22]   Alphonsa Juliet Helina JK, Aswin KI, Viswanathan N. Fabrication and analyzing of Drypetes sepiaria encapsulated chitosan hybrid beads as anticorrosion agent. Mater Today Proc [Internet]. Elsevier Ltd, 2021; 47: 1929–1936. Available from https://doi.org/10.1016/j.matpr.2021.03.714

[23] Kumari S, Tiyyagura HR, Pottathara YB, Sadasivuni KK, Ponnamma D, Douglas TEL, et al. Surface functionalization of chitosan as a coating material for orthopaedic applications: A comprehensive review. Carbohydrate Polymers [Internet]. Elsevier Ltd, 2021; 255: 117487. Available from https://doi.org/10.1016/j.carbpol.2020.117487

[24] Bakshi PS, Selvakumar D, Kadirvelu K, Kumar NS. Chitosan as an environment friendly biomaterial – A review on recent modifications and applications. International Journal of Biological Macromolecules [Internet]. Elsevier B.V., 2020; 150: 1072–1083. Available from https://doi.org/10.1016/j.ijbiomac.2019.10.113

[25] El-Haddad MN. Chitosan as a green inhibitor for copper corrosion in acidic medium. International Journal of Biological Macromolecules [Internet]. Elsevier B.V., 2013; 55: 142–149. Available from http://dx.doi.org/10.1016/j.ijbiomac.2012.12.044

[26] Eduok U, Ohaeri E, Szpunar J. Electrochemical and surface analyses of X70 steel corrosion in simulated acid pickling medium: Effect of poly (N-vinyl imidazole) grafted carboxymethyl chitosan additive. Electrochimica Acta [Internet]. Elsevier Ltd, 2018; 278: 302–312. Available from https://doi.org/10.1016/j.electacta.2018.05.060

[27] Balaji J, Sethuraman MG. Chitosan-doped-hybrid/TiO2 nanocomposite based sol-gel coating for the corrosion resistance of aluminum metal in 3.5% NaCl medium. International Journal of Biological Macromolecules [Internet]. Elsevier B.V., 2017; 104: 1730–1739. Available from http://dx.doi.org/10.1016/j.ijbiomac.2017.03.115

[28] Christopher G, Kulandainathan MA, Harichandran G. Biopolymers nanocomposite for material protection: Enhancement of corrosion protection using waterborne polyurethane nanocomposite coatings. Progress in Organic Coatings [Internet]. Elsevier B.V., 2016; 99: 91–102. Available from http://dx.doi.org/10.1016/j.porgcoat.2016.05.012

[29] Giuliani C, Pascucci M, Riccucci C, Messina E, de Luna SM, Lavorgna M, et al. Chitosan-based coatings for corrosion protection of copper-based alloys: A promising more sustainable approach for cultural heritage applications. Progress in Organic Coatings [Internet]. Elsevier, 2018; 122: 138–146. Available from https://doi.org/10.1016/j.porgcoat.2018.05.002

[30] Ma IAW, Sh A, Ramesh K, Bashir V, Ramesh S, Arof AK. Anticorrosion properties of epoxy-nanochitosan nanocomposite coating. Progress in Organic Coatings [Internet]. Elsevier, 2017; 113: 74–81. Available from http://dx.doi.org/10.1016/j.porgcoat.2017.08.014

[31] Sambyal P, Ruhi G, Dhawan SK, Bisht BMS, Gairola SP. Enhanced anticorrosive properties of tailored poly(aniline-anisidine)/chitosan/SiO2 composite for protection of mild steel in aggressive marine conditions. Progress in Organic Coatings [Internet]. Elsevier, 2018; 3: 203–213. Available from https://doi.org/10.1016/j.porgcoat.2018.02.014

[32] Payra D, Naito M, Fujii Y, Nagao Y. Hydrophobized plant polyphenols: Self-assembly and promising antibacterial, adhesive, and anticorrosion coatings. Chemical Communications Royal Society of Chemistry, 2016; 52: 312–315.

[33] Silva da Conceição DK, Nunes de Almeida K, Nhuch E, Raucci MG, Santillo C, Salzano de Luna M, et al. The synergistic effect of an imidazolium salt and benzotriazole on the protection of bronze surfaces with chitosan-based coatings. Herit Sci [Internet]. Springer International Publishing, 2020; 8: 1–14. Available from https://doi.org/10.1186/s40494-020-00381-4

[34] Menaka R, Subhashini S. Chitosan Schiff base as effective corrosion inhibitor for mild steel in acid medium. Polymer International, 2017; 66: 349–358.

[35] Li M, Xu J, Li R, Wang D, Li T, Yuan M, et al. Simple preparation of aminothiourea-modified chitosan as corrosion inhibitor and heavy metal ion adsorbent. Journal of Colloid and Interface Science [Internet]. Elsevier Inc., 2014; 417: 131–136. Available from http://dx.doi.org/10.1016/j.jcis.2013.11.053

[36] Calegari F, da Silva BC, Tedim J, Ferreira MGS, Berton MAC, Marino CEB. Benzotriazole encapsulation in spray-dried carboxymethylcellulose microspheres for active corrosion protection of carbon steel. Progress in Organic Coatings [Internet]. Elsevier, 2020; 138: 105329. Available from https://doi.org/10.1016/j.porgcoat.2019.105329

[37] Li Z, Kuang H, Yang J, Hu J, Ding B, Sun W, et al. Improving emulsion stability based on ovalbumin-carboxymethyl cellulose complexes with thermal treatment near ovalbumin isoelectric point. Scientific Reports [Internet]. Springer US, 2020; 10: 1–9. Available from http://dx.doi.org/10.1038/s41598-020-60455-y

[38] Alves L, Ferraz E, Gamelas JAF. Composites of nanofibrillated cellulose with clay minerals: A review. Advances in Colloid and Interface Science [Internet]. Elsevier B.V., 2019; 272: 101994. Available from https://doi.org/10.1016/j.cis.2019.101994

[39] Javanbakht S, Shaabani A. Carboxymethyl cellulose-based oral delivery systems. International Journal of Biological Macromolecules [Internet]. Elsevier B.V., 2019; 133: 21–29. Available from https://doi.org/10.1016/j.ijbiomac.2019.04.079

[40] Finkenstadt VL, Bucur CB, Côté GL, Evans KO. Bacterial exopolysaccharides for corrosion resistance on low carbon steel. Journal of Applied Polymer Science, 2017; 134: 1–7.

[41] Hasanin MS, Al Kiey SA. Environmentally benign corrosion inhibitors based on cellulose niacin nano-composite for corrosion of copper in sodium chloride solutions. International Journal of Biological Macromolecules [Internet]. Elsevier B.V., 2020; 161: 345–354. Available from https://doi.org/10.1016/j.ijbiomac.2020.06.040

[42] Znini M, Majidi L, Bouyanzer A, Paolini J, Desjobert JM, Costa J, et al. Essential oil of Salvia aucheri mesatlantica as a green inhibitor for the corrosion of steel in 0.5M H2SO4. Arabian Journal of Chemistry [Internet]. King Saud University, 2012; 5: 467–474. Available from http://dx.doi.org/10.1016/j.arabjc.2010.09.017

[43] Cui G, Bi Z, Wang S, Liu J, Xing X, Li Z, et al. A comprehensive review on smart anti-corrosive coatings. Progress in Organic Coatings [Internet]. Elsevier, 2020; 148: 105821. Available from https://doi.org/10.1016/j.porgcoat.2020.105821

[44] Abbaspoor S, Ashrafi A, Salehi M. Cathodic disbonding of self-healing composite coatings: Effect of ethyl cellulose micro/nanocapsules. Corrosion Engineering, Science and Technology [Internet]. Taylor & Francis, 2021; 56: 659–667. Available from https://doi.org/10.1080/1478422X.2021.1937451

[45] Ji X, Wang W, Zhao X, Zhang B, Chen S, Sun Y, et al. Preparation and properties of self-healing anticorrosive coating on methyl cellulose core-shell fibers. Materials Letters [Internet]. Elsevier B.V., 2021; 290: 129504. Available from https://doi.org/10.1016/j.matlet.2021.129504

[46] Joshy KS, Jose J, Li T, Thomas M, Shankregowda AM, Sreekumaran S, et al. Application of novel zinc oxide reinforced xanthan gum hybrid system for edible coatings. International Journal of Biological Macromolecules [Internet]. Elsevier B.V., 2020; 151: 806–813. Available from https://doi.org/10.1016/j.ijbiomac.2020.02.085

[47] Sushmitha Y, Rao P. Electrochemical investigation on the acid corrosion control of mild steel using biopolymer as an inhibitor. Port Electrochim Acta, 2020; 38: 149–163.

[48] Mousaa IM. Synthesis and performance of bio-based unsaturated oligomer and containing gum arabic as a novel protective steel coating under UV irradiation. Progress in Organic Coatings [Internet]. Elsevier, 2020; 139: 105400. Available from https://doi.org/10.1016/j.porgcoat.2019.105400

[49] Nawaz M, Shakoor RA, Kahraman R, Montemor MF. Cerium oxide loaded with Gum Arabic as environmentally friendly anti-corrosion additive for protection of coated steel. Materials & Design [Internet]. The Author(s), 2021; 198: 109361. Available from https://doi.org/10.1016/j.matdes.2020.109361

[50] Arukalam IO, Ishidi EY, Obasi HC, Madu IO, Ezeani OE, Owen MM. Exploitation of natural gum exudates as green fillers in self-healing corrosion-resistant epoxy coatings. Journal of Polymer Research, 2020; 27: 80.

[51] Sabri S, Najjar A, Manawi Y, Eltai NO, Al-Thani A, Atieh MA, et al. Antibacterial properties of polysulfone membranes blended with Arabic gum. Membranes (Basel), 2019; 9: 1–16.

[52] Solomon MM, Gerengi H, Umoren SA, Essien NB, Essien UB, Kaya E. Gum Arabic-silver nanoparticles composite as a green anticorrosive formulation for steel corrosion in strong acid media. Carbohydrate Polymers [Internet]. Elsevier, 2018; 181: 43–55. Available from https://doi.org/10.1016/j.carbpol.2017.10.051

[53] Verma C, Quraishi MA. Gum Arabic as an environmentally sustainable polymeric anticorrosive material: Recent progresses and future opportunities. International Journal of Biological Macromolecules [Internet]. Elsevier B.V., 2021; 184: 118–134. Available from https://doi.org/10.1016/j.ijbiomac.2021.06.050

[54] Behpour M, Ghoreishi SM, Khayatkashani M, Soltani N. The effect of two oleo-gum resin exudate from Ferula assa-foetida and Dorema ammoniacum on mild steel corrosion in acidic media. Corrosion Science [Internet]. Elsevier Ltd, 2011; 53: 2489–2501. Available from http://dx.doi.org/10.1016/j.corsci.2011.04.005

[55] Kumar A, Rao KM, Han SS. Application of xanthan gum as polysaccharide in tissue engineering: A review. Carbohydrate Polymers [Internet]. Elsevier, 2018; 180: 128–144. Available from http://dx.doi.org/10.1016/j.carbpol.2017.10.009

[56] Abu Elella MH, Goda ES, Gab-Allah MA, Hong SE, Pandit B, Lee S, et al. Xanthan gum-derived materials for applications in environment and eco-friendly materials: A review. J Environ Chem Eng [Internet]. Elsevier Ltd, 2021; 9: 104702. Available from https://doi.org/10.1016/j.jece.2020.104702

[57] Umoren SA, Obot IB, Ebenso EE. Corrosion inhibition of aluminium using exudate gum from Pachylobus edulis in the presence of halide ions in HCl. E-Journal Chem, 2008; 5: 355–364.

[58] Li H, Qiang Y, Zhao W, Zhang S. 2-Mercaptobenzimidazole-inbuilt metal-organic-frameworks modified graphene oxide towards intelligent and excellent anti-corrosion coating. Corrosion Science [Internet]. Elsevier Ltd, 2021; 191: 109715. Available from https://doi.org/10.1016/j.corsci.2021.109715

[59] Fouda AEAS, El-bendary MM, Etaiw SE din H, Maher MM. Structure, characterizations and corrosion inhibition of new coordination polymer based on cadmium azide and nicotinate ligand. Prot Met Phys Chem Surfaces, 2018; 54: 689–699.

[60] Habibiyan A, Ramezanzadeh B, Mahdavian M, Kasaeian M. Facile size and chemistry-controlled synthesis of mussel-inspired bio-polymers based on Polydopamine Nanospheres: Application as eco-friendly corrosion inhibitors for mild steel against aqueous acidic solution. Journal of Molecular Liquids [Internet]. Elsevier B.V., 2020; 298: 111974. Available from https://doi.org/10.1016/j.molliq.2019.111974

[61] Dastpak A, Ansell P, Searle JR, Lundström M, Wilson BP. Biopolymeric anticorrosion coatings from cellulose nanofibrils and colloidal lignin particles. ACS Applied Materials & Interfaces, 2021; 13: 41034–41045.

Elyor Berdimurodov, Abduvali Kholikov, Khamdam Akbarov,
Khasan Berdimuradov, Omar Dagdag, Rajesh Haldhar,
Mohamed Rbaa, Dakeshwar Kumar Verma, Lei Guo

7 Modern testing and analyzing techniques in corrosion

Abstract: Currently, corrosion of metal materials and their protection is an ongoing and interesting topic in engineering science. The selection of analysis methods in the investigation of anticorrosion nature of chemical compounds is a very important task. In this chapter, the modern testing and analyzing methods include a review of electrochemical, surface, quantum chemical and molecular dynamic simulation, and their main characteristics are discussed with relevant examples. The electrochemical actions in the cathodic and anodic regions on the metal surface in the corrosive and inhibited solution are well defined by the electrochemical methods. The surface morphology and chemical composition change on the metal surface are deeply investigated with the surface testing methods. The corrosion and inhibition properties of organic inhibitors are theoretically investigated with the quantum chemical and molecular dynamic simulation testing methods.

Keywords: Corrosion inhibitors, corrosion tests, electrochemical analysis, theoretical analysis, surface morphology

7.1 Introduction

7.1.1 Analysis in corrosion

Metallic materials are a main part of the chemical industry. Their corrosion is and economic and environmental problem [1–5]. Metallic materials are corroded in acidic, saline, alkaline, and other, solutions [6–8]. In these actions, the metal surface reacted with the corrosive ions chemically to form corrosion deposits, which adsorbed on the metal surface and are the reason for destruction. The corrosive ions are easily adsorbed on the metal surface and increase the corrosion rate of the metal surface [9–12]. The metal is ionized on the anodic regions, and the hydrogen is evaluated on the cathodic regions. The corrosion inhibitors are widely used to protect the metal materials from the corrosion processes [13–16]. In this condition, the corrosion inhibitor is added to the corrosion solution; as a result, the corrosion inhibitor is adsorbed on the metal surface and protects the metal from corrosion. These actions are called inhibition processes [17–21].

https://doi.org/10.1515/9783110760583-007

The corrosion and inhibition processes are deeply analyzed with the electro-chemical, surface, quantum chemical, and molecular dynamic simulation analysis methods. The electrochemical frequency modulation (EFM), electrochemical noise (EN), electrochemical impedance spectroscopy (EIS), potentiodynamic polarization (PDP), polarization resistance (LPR), and cyclic polarization (CV) electrochemical testing methods are widely employed to analyze the corrosion and inhibition processes in corrosion science [22–26]. The chemical and physical actions in the cathodic and anodic regions on the metal surface in corrosion and inhibition processes are well defined with the above electrochemical methods [27–30]. The surface analysis methods such as atom force microscopy (AFM), scanning electron microscopy (SEM), 3D profilometry, X-ray photoelectron spectroscopy (XPS), angle-resolved X-ray photo-electron spectroscopy (ARXPS), and energy dispersive X-ray spectrometry (EDX) are employed in the surface morphology analysis to describe the changes on the metal surface after corrosion and inhibition [31–33]. Molecular dynamic (MD) simulation and quantum chemical calculations are used as powerful theoretical methods; they can identify the correlation between the inhibitor structure and inhibition properties [34–37].

7.2 Main part

7.2.1 Electrochemical testing in corrosion

Electrochemical tests are widely employed methods in corrosion analysis. The elec-trochemical kinetics of corrosion and inhibition processes is deeply investigated by the electrochemical techniques. The corrosion and inhibition actions depend on the nature of cathodic and anodic electrochemical processes on the metal interface/elec-trolyte solution. The metal is ionized to form metal ions on the anodic sites [38–41]. In comparison, the hydrogen evolution and some cathodic processes occurred on the cathodic sites. Currently, PDP, EIS, EN, EFM, LPR, and CV are the most used and ef-fective electrochemical testing methods in corrosion analysis. Electrochemical analy-sis is performed in various electrochemical equipment [42–45]. Their installation and electrochemical techniques are similar. Currently, the Gamry electrochemical equip-ment (USA) is widely employed in corrosion testing due to its accuracy in calculation. This equipment gives the raw data of the electrochemical investigation. Then, the Gamry Echem Analyst software is deployed to analyze the raw data of obtained elec-trochemical results. In electrochemical experiments, several types of electrodes are used. For example, the reference electrode is used to compare the obtained electro-chemical data to targeted experiments. The saturated calomel electrode is mostly used as a reference electrode [46–51]. This is due to its low cost and good effective-ness. The working electrode is a selected metal sample; it may be steel, copper, aluminum, or other metal. The selection of metal types of electrochemical research depends on the corrosion material. A dimension exposure area of 1 cm² of the working electrode is widely employed [52–55]. The next important electrode is the counter electrode; it is used to count the electrochemical data more accurately. The platinum electrode is widely performed as a counter electrode as it is an accurate and low resistance electrode [56–58]. Another significant factor is the nature of electrolytes. Corrosion processes have occurred in various electrolyte solutions. The acidic, saline, alkaline, neutral, and other corrosion solutions are mostly used. Before doing the electrochemical experiments, all studied electrodes are immersed for 76 h. This is because electrochemical potential achieves a stable position on the metal surface after immersion time [15, 17].

7.2.1.1 PDP testing in corrosion

Corrosion kinetics is well described by the PDP testing. The corrosion rate, corrosion current density, Tafel slopes and corrosion–inhibition potentials in anodic and cathodic corrosion processes are accurately measured using Tafel polarization. Generally, the PDP test is done between ±0.250 V potential ranges. Before doing PDP analysis, the open-circuit potential (OCP) is calculated. The OCP (Figure 7.1a) can show the chemical–thermodynamic stability of protective film of corrosion inhibitors on the metal electrode. The metal oxide, hydroxyl, and salts are formed on the metal surface from the corrosion processes [25, 26]. When the corrosion inhibitors are added to the corrosion solution, the protective layer of the corrosion inhibitor is formed; as a result, the metal surface is effectively insulated from the aqueous phase of the corrosion solution. It was shown that the stability of corrosion products (metal oxide, hydroxyl, and salts) and protective films on the metal surface are identified from the obtained OCP curves [27]. If the OCP curves are stabilized around negative potential, it indicates that the corrosion product or protective film is more stable. When the OCP curves changed to a more negative or positive potential, it indicates that the corrosion products are not stable. The OCP of inhibition solution shows the inhibitor type. If the obtained OCP changed to a more negative potential, then it indicates that the studied inhibitor is the cathodic type [28]. In comparison, if the obtained OCP changed to a more positive potential, it is vice versa. Tafel polarization curves are also found in PDP testing. They indicate the polarization nature of cathodic and anodic sites in the corrosive and inhibited solution [29]. It is clear from Figure 7.1b that the Tafel curves were cited in the higher corrosion current density regions, showing that the metal surface was importantly corroded and more polarized [30]. The aggressive corrosion ions effectively reacted with the metal surface; as a result, the corrosion on the metal surface was high rate. When the corrosion inhibitor is added to the corrosion solution, Tafel plots are shifted to lower corrosion current density, suggesting that the corrosion actions are blocked by the formation of protective film [31]. The corrosion inhibitors contained many electron-rich sites, which are responsible for the corrosion protection on the metal surface. The delocalized electrons of electron-rich nitrogen atoms, aromatic rings, and functional groups are shared to metal surface. This process is called chemical adsorption mechanism. It is mainly responsible for the inhibition processes. The change of Tafel plots is responsible for the corrosion inhibitor blocking the cathodic and anodic electrochemical processes [32, 33].

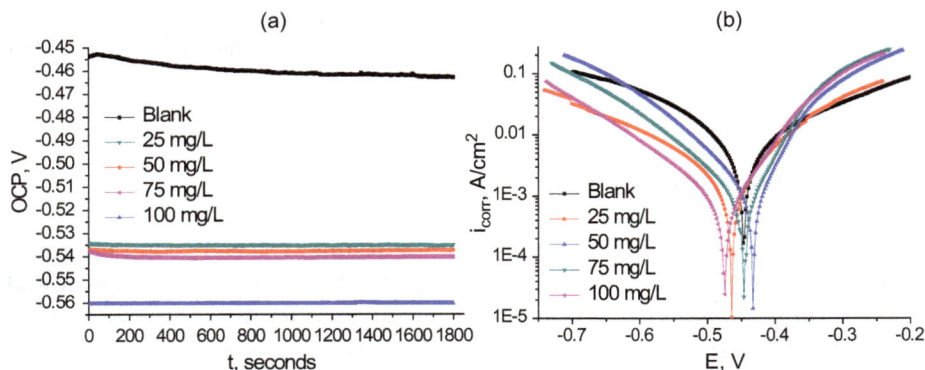

Figure 7.1: (a) OCP and (b) Tafel plots for blank and organic compound (at various concentrations) in acidic solution (1 M HCl) (reprinted with permission from [27], © 2021, Elsevier).

On the anodic sites, the iron reacted with the corrosive chloride ions to form iron salts, which are good adsorbents on the metal surface. Some iron ions are acted in the solution to form free iron ions, which are responsible for the next corrosion processes. These anodic reactions are following [28]:

$$Fe + Cl^- \leftrightarrow FeCl^-_{(ads)}$$

$$FeCl^-_{(ads)} \leftrightarrow FeCl_{(ads)} + e^-$$

$$FeCl_{(ads)} \leftrightarrow FeCl^+_{(ads)} + e^-$$

$$FeCl^+_{(ads)} \leftrightarrow a\,Fe^{2+} + Cl^-$$

In comparison, hydrogen ions are formed in the cathodic regions. The hydrogen ions are also combined to form hydrogen gas. At end of the cathodic processes, the hydrogen gas was evaluated. These cathodic reactions are following [29]:

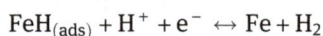

$$Fe + H^+ \leftrightarrow FeH^+_{(ads)}$$

$$FeH^+_{(ads)} + e^- \leftrightarrow FeH_{(ads)}$$

$$FeH_{(ads)} + H^+ + e^- \leftrightarrow Fe + H_2$$

In PDP testing, the values in corrosion rate, (CR$_{PDP}$), Tafel anodic beta, cathodic beta slopes, corrosion potential, (E_{corr}), and corrosion current density, (i_{corr}), for corrosion and inhibition processes are calculated. The values of corrosion current density are performed to measure the protection degree of corrosion inhibitor, (η_{PDP}) as follows:

$$\eta_P, \% = \frac{i^o_{PDP} - i^i_{PDP}}{i^o_{PDP}} \times 100\% \tag{7.1}$$

where i^o_{PDP} and i^i_{PDP} are the corrosion currents in the corrosion and inhibition processes, respectively.

Tafel polarization curves are employed to estimate the electrochemical data. The electrochemical kinetics are described using the values in the corrosion rate (CR$_{PDP}$), Tafel anodic beta, cathodic beta slopes, corrosion potential (E_{corr}), and corrosion current density (i_{corr}). For example, the values of CR$_{PDP}$ and i_{corr} are very high in the corrosion solution [30]. When the corrosion inhibitor is introduced into the corrosion solution, the values of CR$_{PDP}$ and i_{corr} decrease, indicating that the hydrogen evolution and iron dissolution are reduced. The corrosion potential difference between the inhibited and uninhibited solution identifies the type of corrosion inhibitor [31]. When this potential gap is over 80 mV, then the inhibitor is classified as cathodic. In contrast, if this potential gap is under 80 mV, then the inhibitor is an anodic type. The changes in Tafel anodic beta and cathodic beta slopes describe the inhibitor impacts to the cathodic and anodic electrochemical actions [32].

7.2.1.2 EIS testing in corrosion

EIS method is a very efficient method in corrosion research. Many important electrochemical parameters are estimated using EIS. Generally, the EIS investigation is performed with the following requirements: 10 mV$_{rms}$ AC voltage amplitude, 0.01 Hz– 100 kHz frequencies. Bode and Nyquist plots are found in EIS investigation. Nyquist plots can describe the nature of the charge transfer mechanism. For example, Figure 7.2a indicates the Nyquist plots for a metal electrode in the corrosion and inhibited solution. It is clear from Figure 7.2a that the resulting polarization curves were depressed semi-rings when the inhibitor was present or absent in the medium, proving that the metal's destruction depends on the charge-sharing actions between the anode and cathode [33, 34]. These plots are smaller in the corrosion system than in the inhibited solution. This is due to the aggressive corrosive ions polarizing the metal surface. The size of polarization curves increases with the increase of concentration, showing that the electrochemical kinetics depends on the change of concentrations. The next important factor is that these impedance plots are cited at higher frequencies in the inhibited solution, suggesting that the protective film covered the metal surface and decreased the metal polarization [35, 36].

Figure 7.2b indicated that the phase angle and Bode plots are cited at a higher frequency, showing that the metal/electrolyte interface exhibited nonideal capacitance nature. Bode plots are one-time constant at medium frequency, demonstrating that the relaxation processes have occurred on the corrosion and inhibition processes [37, 38].

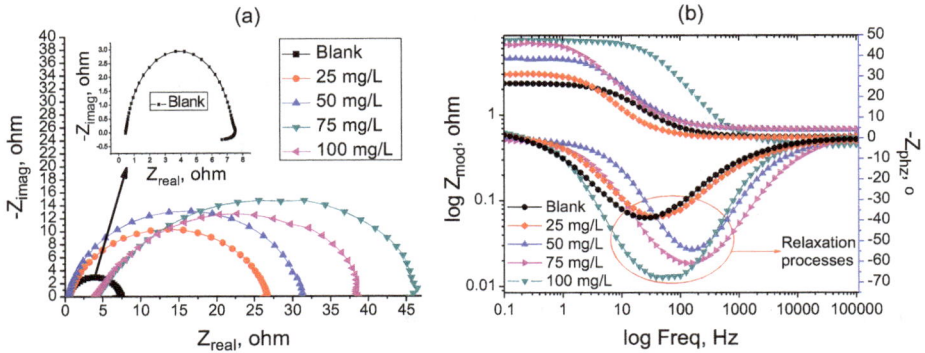

Figure 7.2: (a) Nyquist, (b) phase angle–Bode plots in the corrosive and inhibited solution with various inhibitor concentrations (reprinted with permission from [24], © 2021 Elsevier).

In the calculation of EIS data, the equivalent circuit model is selected. The selection of an equivalent circuit model depends on the research type. Nyquist plots are used to estimate the data of EIS. Generally, it includes CPE (constant phase element), R_{ct} (charge transfer resistance), and R_s (solution resistance). CPE has been performed for the capacitor to accurately fit the circuit, which is specified shown in eq. (7.2). In this expression, the empirical constant (n), CPE magnitude (Y_0), CPE imaging number (i), and angular frequency (ω) were used to estimate the capacitor [5, 7]:

$$Z_{CPE} = Y_0^{-1}(i\omega)^{-n} \tag{7.2}$$

In EIS testing for corrosion, the values in solution resistance (R_s), protective layer resistance (R_{pl}), charge transfer resistance (R_{ct}), diffusion layer resistance (R_{dl}), accumulation resistance (R_a), and double-layer capacitance, (C_{dl}) are estimated. The obtained values of R_{ct} are used to calculate the inhibition efficiency, ($\eta_{EIS}, \%$), as follows [6, 9]:

$$\eta_{EIS}, \% = \frac{R_{ct}^i - R_{ct}^o}{R_{ct}^i} \times 100\% \tag{7.3}$$

where R_{ct}^i and R_{ct}^o are the charge transfer resistances in the solutions where the inhibitor is present or absent, respectively.

The local dielectric constant is described using the obtained amounts of double-layer capacitance from the Helmholtz model (eq. (7.4)). It can show that the inhibitor substituted the pre-adsorbed water molecules on the metal surface. As a result, the electrical conductivity of the solution decreases (eq. (7.5)) [11, 12]. In this expression, the empirical constant (A), selected medium's dielectric constant (ε^0), double layer thickness (d), and vacuum's dielectric constant (ε), were applied to estimate the double-layer capacitance:

$$C_{dl} = (Y_0 R_{ct}^{1-n})^{1/n} \tag{7.4}$$

$$C_{dl} = \frac{\varepsilon\varepsilon^0}{d} A \tag{7.5}$$

7.2.1.3 EN testing in corrosion

Currently, EN testing is a new technique in corrosion research. It is a nondestructive and sensitive research method in corrosion and inhibition investigation. In EN, the statistical time versus current and potential change is employed to estimate the EN data. The current and potential noise is calculated from the following equations [31, 36]:

$$I_{bf}[n] = A_i T[n] + B_i \tag{7.6}$$

$$E_{bf}[n] = A_e T[n] + B_e \tag{7.7}$$

In these equations, E_{bf} is the potential of standard linear best fit; B is the Stern–Geary constant; T is the time at the start of the data block, and n is the total point number; A is an anodic coefficient; I_{bf} is the current of standard linear best fit.

Equations (7.8) and (7.9) are named standard linear best fitting; they can illustrate the dependence of time to potential and current; they are also used to estimate the DC trend. Equations (7.8) and (7.9) are performed to measure the RMS (root means square) of the residual AC signal, and the electrochemical noise of block is calculated by subtracting the DC trend [15, 32]:

$$I_{rms} = \sqrt{\frac{\sum_{n=1}^{N} (I[n] - I_{bf}[n])^2}{N}} \tag{7.8}$$

$$E_{rms} = \sqrt{\frac{\sum_{n=1}^{N} (E[n] - E_{bf}[n])^2}{N}} \tag{7.9}$$

where n is the total point number; N is the total channel number; E_{bf} is the potential of standard linear best fit; I_{bf} is the current of standard linear best fit; I_{rms} is the current noise; and E_{rms} is the potential noise.

In the EN experiment, the potential and current noise versus time plots are found. Next, the average values in EN noise resistance are calculated to use the average values of the potential ($\bar{E}_{rms}$) and current ($\bar{I}_{rms}$) noise in the corrosion and inhibited solution are estimated according to the following equation [6, 9]:

$$\bar{R}_{rms} = \frac{\bar{E}_{rms}}{\bar{I}_{rms}} \qquad (7.10)$$

Table 7.1 describes the estimated EN parameters: $\bar{I}_{rms}$, $\bar{V}_{rms}$, $\bar{R}_{rms}$, IE_{EN}, and q (total charge). The average noise resistance is larger in corrosive solution than that in the inhibited solution, showing that the corrosion inhibitor effectively increases the electrochemical noise resistance. $\bar{R}_{rms}$ is used to measure the protection degree of corrosion inhibitor as follows [8, 10]:

$$IE_{EN}, \% = \frac{\bar{R}_{rms}^{inh} - \bar{R}_{rms}^{0}}{\bar{R}_{rms}^{inh}} \times 100 \qquad (7.11)$$

In this equation, IE_{EN} is the inhibition protection degree; $\bar{R}_{rms}^{inh}$ is the average noise resistance in inhibited solution; $\bar{R}_{rms}^{0}$ is the average noise resistance in corrosion solution.

The values of total charge indicate the charge transfer between the anode and cathode sites. When the charge transfer is high, the values of the total charge become high; it is an indicator of the high rate of corrosion. When the corrosion inhibitor is added to the corrosion solution, the total charge values are decreased because the charge transfer action is depleted in the cathodic and anodic sites on the metal surface with the formation of the protective film [3, 6].

Table 7.1: EN data for corrosion inhibitor (reprinted with permission from [29], © 2021 Elsevier).

C_{inh}, mg/L	$\bar{I}_{rms}$, µA	$\bar{V}_{rms}$, µV	$\bar{R}_{rms}$, Ω	IE_{EN}, %	q, C
Blank	9.69	382	39.42	–	4.886×10^{-2}
50	37.91	7,540	198.87	80.17	2.95×10^{-5}
100	73.792	21,430	290.41	86.42	5.604×10^{-5}
150	74	29,500	398.65	90.11	11.04×10^{-5}
200	107.646	48,470	450.27	91.24	23×10^{-5}

7.2.1.4 EFM, LPR and CV testing in corrosion

In corrosion research, the EFM, LPR, and CV testing methods are also widely employed to describe the electrochemical processes on the metal/electrolyte system. For example, EFM is a faster and sensitive testing method in corrosion. The corrosion rate and Tafel slopes are easily estimated using the EFM method. The inhibition efficiency of corrosion inhibitors is easily calculated. The half- electrochemical reactions of cathodic and anodic regions on the metal surface in the absence and presence of inhibitor solutions are well organized [6]. LPR is the next more sensitive method in corrosion analysis. The value in the polarization resistance is calculated accurately by LPR. The inhibition efficiency is calculated using the values in the linear polarization resistance. LPR research shows that the polarization resistance of metal electrodes in the aggressive corrosive medium is very low compared to that in the protective system, because the protective film on the metal surface increases the polarization resistance [14]. The CV method indicates the changes in reduction and oxidation potential on the metal surface in the corrosion and inhibition processes. The inhibitor addition shifts the values in the oxidation and reduction potential.

7.2.2 Surface analysis in corrosion

Surface analysis is the next important step in investigating the surface phenomenon of metal samples after corrosion and inhibited processes. The corrosion destruction and inhibition protection of metal surface is illustrated by surface analysis. The surface morphology on the metal surface is investigated using AFM, SEM, 3D profilometry, XPS, ARXPS, and EDX methods. Before surface analysis, the metal samples are immersed for 1–3 h in the studied solutions. Next, the acetone and distilled water are used to wash the immersed metal sample. Then, the dried metal sample is degreased with acetone [16, 17].

7.2.2.1 SEM testing in corrosion

SEM tests show the surface destruction of metal after the corrosion and inhibited processes. The micrographs of the metal surface for the inhibited and corrosion solution are obtained in SEM analysis. Figure 7.3a shows that the metal surface is seriously corroded and destructed in 1 M NaCl + 1 M NaCl solutions after 3 days of immersion time. This is due to the metal surface being destructed under the corrosive ions. More corrosive deposits (iron hydrochlorides, hydroxides, and salts) are formed on the metal surface [18, 23]. On the other hand, the surface morphology is significantly improved with the addition of corrosion inhibitors (Figure 7.3b) because the metal surface is protected with the thin film of corrosion inhibitor. The electron-rich nitrogen and oxygen atoms in the inhibitor structure are responsible for the formation of protective film on the metal surface. The free electrons are transferred to the vacant d-orbitals of metal; consequently, covalent bonds are formed between the metal surface and corrosion inhibitor. Therefore, SEM investigation results confirm the formation of protective film that can effectively insulate the metal surface from the corrosive solution [24, 26].

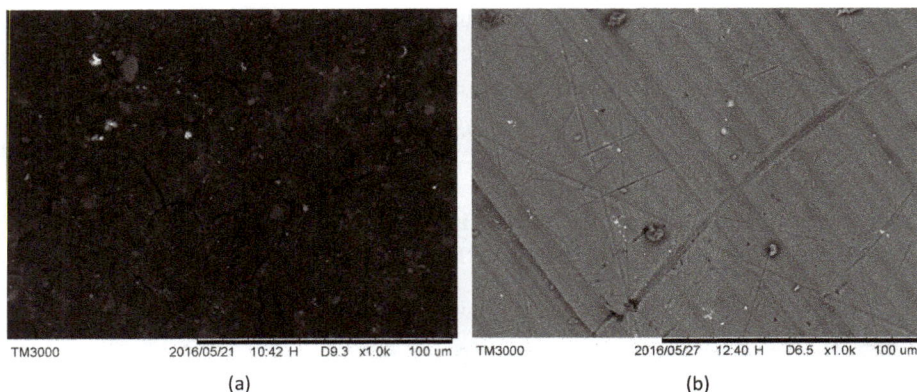

Figure 7.3: SEM micrographs for carbon steel in (a) the inhibitor absence and (b) inhibitor presence 1 M NaCl + 1 M NaCl solutions after 3 days of immersion (reprinted with permission from [20], © 2021 Elsevier).

7.2.2.2 EDX testing in corrosion

EDX method indicates the elemental compositions of the metal surface after the corrosion and inhibition. For example, Figure 7.4 illustrates EDX results for corroded and inhibited metal sample. It is evident from Figure 7.4a that the oxygen and chloride spectra are high for the corroded surface. This is because metal oxides and chloride salts have covered the metal surface. The corrosion solution is HCl, which reacted with the metal surface to form chloride salt. Then, the salts (corrosion products) cover the metal surface [16]. In comparison, Figure 7.4b reveals that the chloride and oxygen spectra decreased; the sulfur and nitrogen signals have appeared

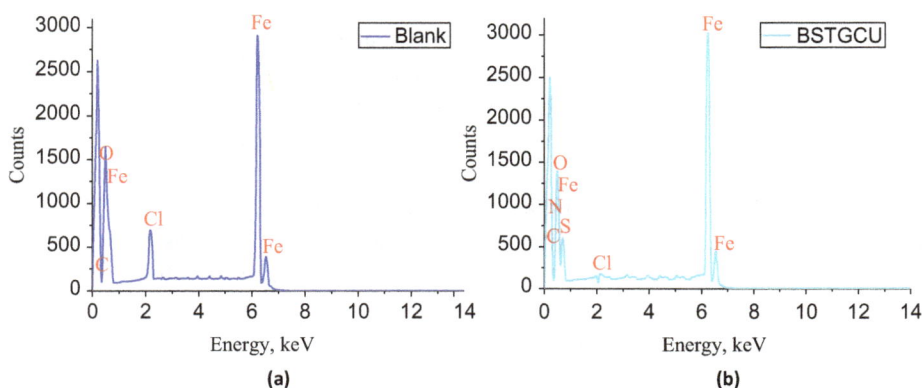

Figure 7.4: EDX micrographs for (a) blank and (b) corrosion inhibitor (reprinted with permission from [27], © 2021 Elsevier).

in inhibited metal surfaces. This is responsible for the decreased amounts of corrosion products in the presence of corrosion inhibitors. The sulfur and nitrogen atoms are mainly responsible for the corrosion protection and formation of protective film on the metal surface. Finally, the EDX test is an important technique to describe surface morphology [25, 28].

7.2.2.3 3D profilometry and AFM testing in corrosion

The surface roughness of the metal sample after corrosion and inhibition processes is analyzed with 3D profilometry and AFM testing. Two roughness levels are named: R_q (the root mean square average of the roughness profile ordinates or root mean square roughness) and R_a (the arithmetic average of the absolute values of the roughness profile ordinates or mean roughness). Topography and morphology investigations are also carried out in 3D profilometry and AFM testing [32, 36]. The 3D profilometry gives 1–100 µm size micrographs, and AFM gives 1–1,000 nm size pictures of the metal surface. The homogeneity of the protective film is also identified with these surface analysis methods. The results of 3D profilometry and AFM testing indicate that the surface roughness is very low on the metal surface after corrosion. The corrosion deposits are formed on the metal surface; as a result, the surface roughness is lessened. In comparison, the corrosion inhibitor formed an active complex on the metal surface; this complex is responsible for high surface roughness on the metal surface [21, 26].

7.2.2.4 XPS testing in corrosion

XPS is used in corrosion research to analyze the structural and deep surface morphology of the metal surface. Currently, several XPS types of equipment are employed. Kratos (Manchester, United Kingdom) and ThermoFisher Scientific K–Alpha Spectrometer (USA) are widely used to do EXP tests. In EXP research, basically, Al K_α X-ray source (1,487 eV) and a monochromator are used. The analyses of the photoelectron spectra on different regions of the surface of the sample were collected and stored in an interfaced PC, using Thermo Avantage V5.938 software. The 400 µm spot size X-ray beam was set in two-pass energy modes of 200 and 20 eV, respectively. The vacuum in the analyzing chamber was kept at 10–9 mbar during data collection. The binding energy (EB) scale for the MePhI-treated brass samples was corrected using the C–C/C–H peak centered at 284.8 eV in the C 1s spectrum for the spectra with a clearly developed C 1s spectrum. The instrument had a hemispherical analyzer that operated at pass energy of 160 eV for a survey and 20 eV for high-resolution (HR) XPS spectra [4, 6, 18].

XPS analysis shows the spectra of related chemical elements, the chemical structure of the surface layer, chemical bonds, and compositions. It is noted in XPS results that the metal ions spectra are highly sensitive in the XPS spectra for the corrosive solution than the neutral metal, confirming that the metal cations are formed under the corrosive solution and adsorbed on the metal surface. In comparison, the neutral metal ions' spectra appeared in XPS spectra for the inhibited solution, because, the corrosion inhibitors neutralized metal cations to form the inhibitor metal complex with the metal ions [16, 27]. In addition to this, the inhibitor molecules are covered on the metal surface. On the other hand, the heteroatoms such as nitrogen, sulfur, oxygen, and other elements are mainly responsible for corrosion inhibition. It is clear from obtained XPS results that these heteroatoms spectra also appear in the resulting XPS spectra, confirming that the heteroatoms are adsorbed on the metal surface. These atoms are also adsorption centers of corrosion inhibitors, which means that the corrosion inhibitor is adsorbed through the adsorption centers [28, 31].

7.2.2.5 Angle-resolved X-ray photoelectron spectroscopy (ARXPS) testing in corrosion

ARXPS testing is also the next important method in corrosion research. This method is carried out at six angles: 10°, 20°, 27°, 35°, 50°, and 90°. ESCApe software is used to control the ARXPS data. The various Ar–ion cluster energies and sizes are used with low energy per atom cluster (5 keV $Ar_{2,000}^+$, 15 s sputtering cycle), followed by 5 keV $Ar_{1,000}^+$ (120 s sputtering cycle), and 10 keV $Ar_{1,000}^+$ (120 s sputtering cycle). The profiling analysis was completed with monoatomic 5 keV Ar^+ (120 s sputtering cycle). The C 1s spectrum signals on ARXPS data indicate the C–C/C–H–containing compounds adsorption on the metal surface. Mostly, the C 1s spectrum appeared on the top spectrum with 10° angle (around 280 eV). The N 1s spectra of ARXPS illustrate C–N bonds on the metal surface. In comparison, the O 1s and Cl 2p signals of ARXPS are responsible for the metal oxides and chloride salts on the metal surface. ARXPS spectra also show the metal signals such Cu 2p, Zn 2p, and Fe 3p. The organic–metallic connection (between metal and corrosion inhibitor) is also confirmed in the ARXPS analysis. The metal–N, P, O and other heteroatoms bonds give signals on the ARXPS spectra. In addition to this, the metal clusters with the corrosion inhibitor or without inhibitor on the metal surface are confirmed by ARXPS methods [21, 23, 31].

7.2.3 Theoretical analysis in corrosion

7.2.3.1 MD simulation testing in corrosion

The interaction energy between the corrosion inhibitor and metal surface, the visualizing of corrosion and inhibition, the adsorption position, and mechanism are theoretically investigated using MD simulation. The MD simulation is carried out mostly in Materials Studio software (BIOVIA, USA, Version 8.0,) with the Forcite module. In MD simulation, analysis was performed using following methods and techniques: (i) Berendsen thermostat is used to set temperature at 298 K; (ii) a simulation time (1 fs to 500 ps), NVT ensemble, and COMPASS II method were used to estimate the theoretical adsorption model on the Fe9110 [26, 28].

The bulk metal crystals are used as a simulation sample. They are cleaved in metal crystallographic planes with various thicknesses. Smart algorithm is used to relax the cleaved surface. Next, a vacuum slab with zero thickness is subsequently built above the supercell the periodicity and relaxed surface is modified through building a 12×10 supercell. In the next step of MD simulation, the molecular structure of the corrosion inhibitor is drawn and the obtained structure is optimized. Next, the simulation box is selected. The size of the simulation box is very important; it needs to contain the corrosion inhibitor, solution molecules, and some ions. The simulation box needs to be simulated with the concentration of corrosive ions (HCl, NaCl, $Ca(OH)_2$, H_2SO_4 and others) [59–62].

The energy interaction ($E_{interaction}$) between the corrosion inhibitor and metal surface is theoretically estimated as shown in eq. (7.12). MD simulation results suggested that when the values of $E_{interaction}$ are high for selected corrosion inhibitor, this molecule is strongly adsorbed on the metal surface [63–65]:

$$E_{interaction} = E_{total} - \left(E_{surface+solution} + E_{inhibitor}\right) \tag{7.12}$$

where $E_{inhibitor}$ is the energy of corrosion inhibitor; $E_{surface+solution}$ is the energy of the simulated metal surface with the solution; E_{total} is the total energy of the studied system [66–68].

The equilibrium configuration (Figure 7.5a), density field distribution (Figure 7.5b), relative concentration (Figure 7.5c), and RDF plot (Figure 7.5d) for the 6-aminopenicillanic acid sodium gossypol (APASG)/Fe(110) system indicate the MD simulation results. When equilibrium was achieved between the temperature and system energy, MD simulation analysis was performed. Figure 7.5a indicates the equilibrium configuration of side and top views for investigated corrosion inhibitors. It is clear from Figure 7.5a that the corrosion inhibitor is adsorbed on the metal surface by the parallel position. The adsorption centers are mainly responsible for adsorption. When the inhibitor concentration increases, the adsorption position of the inhibitor is changed to the vertical position. The inhibitor is maximally covered through the parallel adsorption position [28, 29].

Figure 7.5b illustrates the distribution adsorption field (DAF) of corrosion inhibitor. This MD simulation result confirmed that the hydrophobic barrier and dense are formed to insulate the metal from the electrolyte. The formation of hydrophobic and dense barrier effectively insulate the metal from the corrosion solution. It is important to note MD simulation results in Figure 7.5c that the distance between the organic molecule and Fe(110) surface is 2.836 Å [12, 16].

Figure 7.5d explains the radial distribution function (RDF) and g(r). The average value of g(r) is responsible for the average length of bonds between the corrosion inhibitor and metal surface. The peaks higher than 3.5 Å in the radial distribution function (RDF) and g(r) suggest the physisorption nature of organic compound on Fe(110) surface. In comparison, the peaks between 1 and 3.5 Å in the RDF and g(r) are responsible for the chemisorption nature of corrosion inhibitor. Therefore, the MD simulation results theoretically confirm the electrochemical and surface analysis results. The bond length between the metal and corrosion inhibitor is also found by MD simulation analysis. It can be seen from Figure 7.5d that the bond length between the organic compound and Fe(110) surface is 2.805 Å. It is noted that when the bond length between the metal and corrosion inhibitor is under 3.5 Å, the inhibitor is effectively adsorbed on the metal surface. These details confirmed other research results [24].

The adsorption energy is estimated in the MD simulation. Equation (7.13) is used to measure the average values of adsorption energy [69–72]:

$$E_{ads} = E_{total} - (E_{surf+solu} + E_{inh+solu}) + E_{solu} \tag{7.13}$$

where E_{total} is the total potential energy of the system, which contains the solution, adsorbed inhibitor molecule, and metal crystal; $E_{surf+solu}$ shows the total energy of the metal surface and solution without the inhibitor; $E_{inh+solu}$ indicates the total energy of the inhibitor and solution; E_{solu} is the potential energy of the solvent molecules; and E_{ads} is the adsorption energy [24].

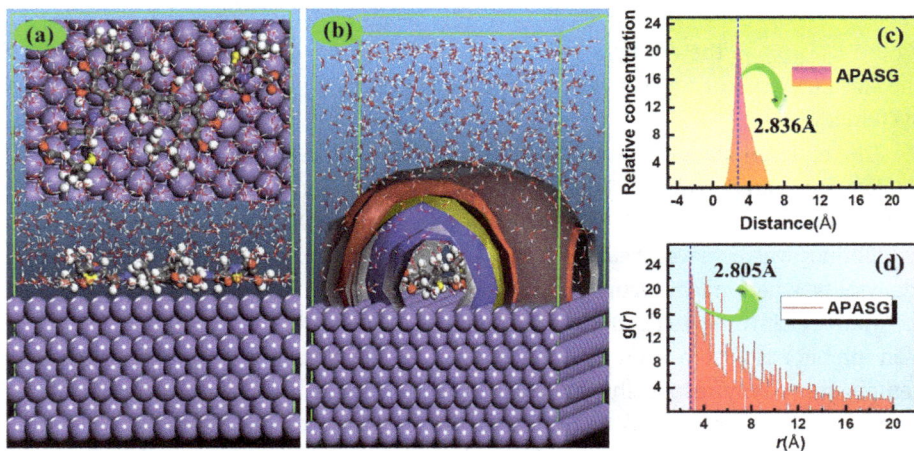

Figure 7.5: **(a)** Adsorption model, **(b)** DAF, **(c)** relative concentration, and **(d)** RDF plot for the corrosion inhibitor (110) system (reprinted with permission from [24], © 2021 Elsevier).

7.2.3.2 Quantum chemical analysis in corrosion

Quantum chemical calculation is the next important tool in corrosion research. The correlation between experimental and theoretical research is found by the quantum chemical calculation. The corrosion and inhibition properties of the metal surface are theoretically investigated in this analysis. Various quantum chemical methods are performed in this analysis. The density functional theory (DFT) with (the three – parameter Lee–Yang–Parr correlation function by Becke) B3LYP method is mostly used in corrosion, among all quantum chemical methods. Different types of basis sets are widely used such as 6-31G, 6-31G (d, p), and 6-311G (d, p) [73–75]. Quantum chemical calculations of selected corrosion inhibitors are estimated in the gas and aqueous phases with the neutral and protonated forms. When the DFT calculation is done in the aqueous phase, a water solvent (the dielectric constant is 80.4, and the refractive index was 1.33) and conductor–like polarizable continuum model (C–PCM) are widely used. The Gaussian is a widely deployed software for quantum chemical calculations [5, 16].

At the start of the DFT calculation, the selected corrosion inhibitor is optimized. The optimized structures indicate the electronic map of molecular structure. It is shown in DFT results that the MEP–LUMO (Figure 7.6f), MEP–HOMO (Figure 7.6e), molecular electrostatic potential (MEP) (Figure 7.6d), lowest molecular orbital (LUMO) (Figure 7.6c), highest occupied molecular orbital (HOMO) (Figure 7.6b), and optimized molecular structure (Figure 7.6a), and for the protonated form of corrosion inhibitor in the aqueous phase, reveal the correlation between the inhibition performance and molecular structure. These results suggest that the studied structure is large and planar, which included the large aminopenicillanic and gossypolone rings [75]. The broadly stretched linked atoms (nitrogen and sulfur) and functional groups (hydroxyl and carboxyl) are mainly responsible for the inhibition performance. The nucleophilic and electrophilic attacks positions of corrosion inhibitors are described in the MEP results. As can be shown, the nitrogen, oxygen and sulfur atoms, benzoyl aromatic rings, and hydroxyl functional groups are attributed to the rise in the nucleophilic feature of organic compound. In comparison, the aldehyde, carboxyl, and methyl electron-rich regions are attributed to an increase in the electrophilic reactivity of the inhibitor.

The obtained HOMO and LUMO show the energy distribution of electron density of the studied molecular structure. The HOMO regions are mainly responsible for the electron donation ability of corrosion inhibitors. The p-electrons of heteroatoms are shared through the HOMO regions. The covalent bonds are formed by the HOMO regions. In comparison, the electron acceptor nature of the corrosion inhibitor promotes electrostatic interaction between the corrosion inhibitor and metal surface. The negatively charged metal surface electrostatically interacts with the positively charged LUMO regions of corrosion inhibitors [24, 27].

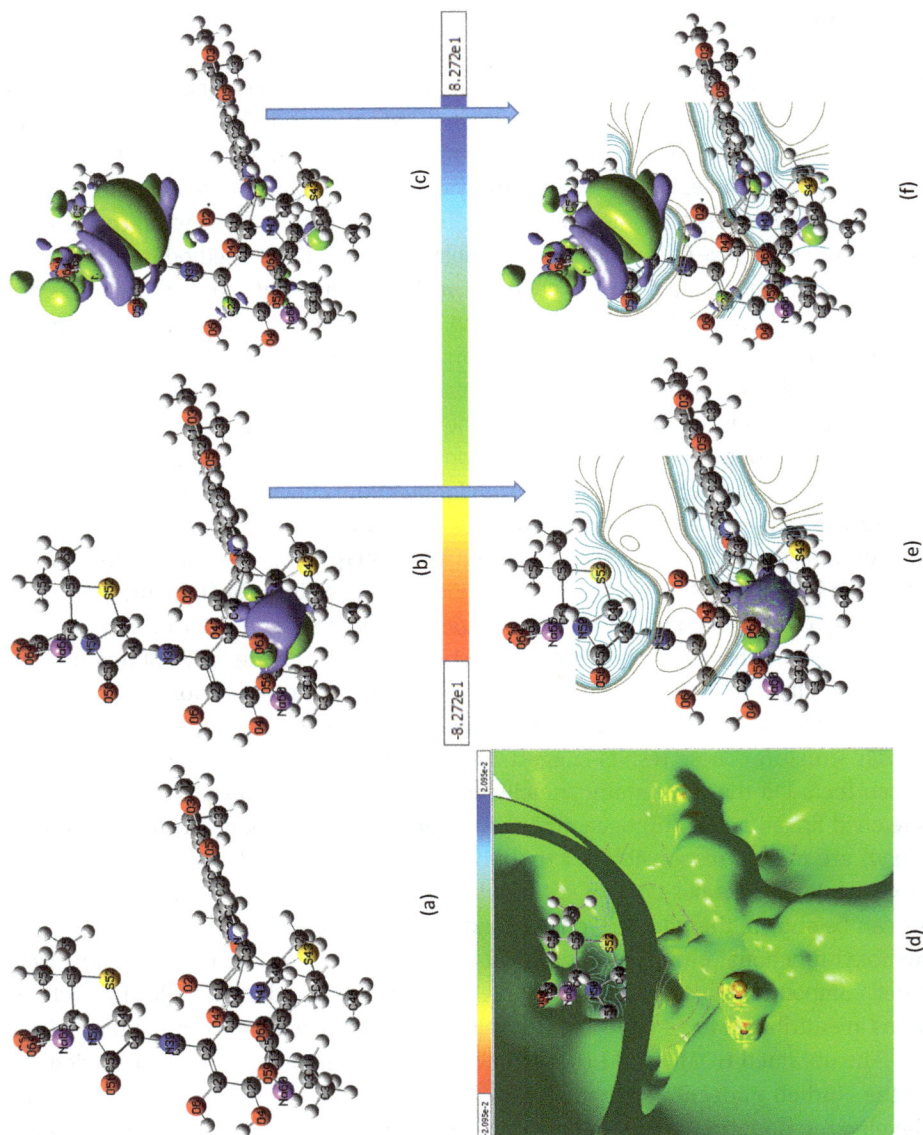

Figure 7.6: (a) Optimized molecular structure, (b) highest molecular occupied orbitals, (c) lowest unoccupied molecular orbitals, (d) MEP, (e) MEP–highest molecular occupied orbitals and (f) molecular electrostatic potential–lowest unoccupied molecular orbitals for the protonated form of corrosion inhibitor in the aqueous phase (reprinted with permission from [24], © 2021 Elsevier).

Local molecular analysis is the next important technique in quantum chemical analysis. The adsorption centers of inhibitor structure are found by the local molecular analysis. The most electrophilic (f_k^+ high), nucleophilic (f_k^- high), and neutral

(f_k^0 high) attacking sites in the selected inhibitor structure were studied by the Fukui function. The Fukui indices (f_k^+, f_k^-, f_k^0) are estimated by the Mulliken population analysis (NPA) as shown in the following equations:

$$f_k^+ = P_k(N+1) - P_k(N) \tag{7.14}$$

$$f_k^- = P_k(N) - P_k(N-1) \tag{7.15}$$

$$f_k^0 = P_k(N+1) - P_k(N-1) \tag{7.16}$$

where $P_k(N+1)$, $P_k(N)$, and $P_k(N-1)$ are anionic, neutral, and cationic of the molecule, respectively.

The global reactive descriptors testing of quantum chemical calculations is also a significant method in corrosion. The energy difference between the HOMO ($E_{HOMO(Inh)}^{DFT}$) and LUMO ($E_{LUMO(Inh)}^{DFT}$) of a corrosion inhibitor is used to estimate the values of global reactive descriptors such as the electron affinity (A_{Inh}^{DFT}), molecular ionization potential (I_{Inh}^{DFT}), electronic negativity (χ_{Inh}^{DFT}), electronic chemical potential (μ_{Inh}^{DFT}), chemical softness (σ_{Inh}^{DFT}), chemical hardness (η_{Inh}^{DFT}), nucleophilicity (ε_{Inh}^{DFT}), and global electrophilicity index (ω_{Inh}^{DFT}), and the estimated values in the electron sharing between the organic compound and metal surface (ΔN_{Inh}^{DFT}) are as shown in the following equations [14, 18, 25]:

$$\Delta E_{Inh}^{DFT} = E_{LUMO(Inh)}^{DFT} - E_{HOMO(Inh)}^{DFT} \tag{7.17}$$

$$A_{Inh}^{DFT} = - E_{LUMO(Inh)}^{DFT} \tag{7.18}$$

$$I_{Inh}^{DFT} = - E_{HOMO(Inh)}^{DFT} \tag{7.19}$$

$$\sigma_{Inh}^{DFT} = \frac{1}{\eta} \tag{7.20}$$

$$\varepsilon_{Inh}^{DFT} = \frac{1}{\omega_{Inh}^{DFT}} \tag{7.21}$$

$$\omega_{Inh}^{DFT} = \frac{(\chi_{Inh}^{DFT})^2}{2\eta_{Inh}^{DFT}} \tag{7.22}$$

$$\eta_{Inh}^{DFT} = \frac{1}{2}(I_{Inh}^{DFT} - A_{Inh}^{DFT}) \tag{7.23}$$

$$-\mu_{Inh}^{DFT} = \chi_{Inh}^{DFT} = \frac{1}{2}(I_{Inh}^{DFT} + A_{Inh}^{DFT}) \tag{7.24}$$

$$\Delta N_{Inh}^{DFT} = \frac{(\chi_{Fe} - \chi_{Inh}^{DFT})}{2(\eta_{Fe} + \eta_{Inh}^{DFT})} \tag{7.25}$$

where η_{Fe} is 0 eV/mol, and χ_{Fe} is 7 eV/mol.

The resulting global reactive descriptors are indicated in Table 7.2. The values $E_{\text{HOMO(Inh)}}^{\text{DFT}}$ and $E_{\text{LUMO(Inh)}}^{\text{DFT}}$ are responsible for the energy of donor and acceptor ability. $\mu_{\text{Inh}}^{\text{DFT}}$ is the electrochemical potential of corrosion inhibitors; it shows that the electrons are easily shared to the metal surface. The values, $\omega_{\text{Inh}}^{\text{DFT}}$, $\varepsilon_{\text{Inh}}^{\text{DFT}}$, $\chi_{\text{Inh}}^{\text{DFT}}$, $\mu_{\text{Inh}}^{\text{DFT}}$, $I_{\text{Inh}}^{\text{DFT}}$, and $A_{\text{Inh}}^{\text{DFT}}$ indicate the electronegativity, good nucleophilic, and low electrophilic nature of the corrosion inhibitor. $\eta_{\text{Inh}}^{\text{DFT}}$ and $\sigma_{\text{Inh}}^{\text{DFT}}$ show the chemical hardness and softness molecules. $\Delta N_{\text{Inh}}^{\text{DFT}} > 0$ indicates that the protection ability of selected compound increases [36, 37]. The values of $\Delta N_{\text{Inh}}^{\text{DFT}}$ accounted for 714.14 eV, indicating that the studied corrosion inhibitor is an excellent defender of metal surface. The values of $\Delta N_{\text{Inh}}^{\text{DFT}}$ indicate the electron-donating nature of the inhibitor. The values in dipole moment show the polarizing degree of the inhibitor. If this value is high, then the inhibitor is more water soluble and is an effective corrosion inhibitor.

Table 7.2: Important quantum chemical parameters of optimized inhibitor structure (reprinted with permission from [24], © 2021 Elsevier).

Parameter	Value, eV
$E_{\text{HOMO(Inh)}}^{\text{DFT}}$	−0.09911
$E_{\text{LUMO(Inh)}}^{\text{DFT}}$	−0.08944
$\Delta E_{\text{Inh}}^{\text{DFT}}$	0.00967
$I_{\text{Inh}}^{\text{DFT}}$	0.09911
$A_{\text{Inh}}^{\text{DFT}}$	0.08944
$\mu_{\text{Inh}}^{\text{DFT}}$	−0.094275
$\chi_{\text{Inh}}^{\text{DFT}}$	0.094275
$\eta_{\text{Inh}}^{\text{DFT}}$	0.004835
$\sigma_{\text{Inh}}^{\text{DFT}}$	206.825
$\omega_{\text{Inh}}^{\text{DFT}}$	0.9191
$\varepsilon_{\text{Inh}}^{\text{DFT}}$	1.088
$\Delta N_{\text{Inh}}^{\text{DFT}}$	714.14
E(RB3LYP), Hartree	−4,023.705315
Dipole moment, D	9.049421

7.3 Conclusion

In this chapter, the main properties and techniques of modern testing and analysis methods were reviewed and discussed. The electrochemical testing methods included PDP, EIS, EN, EFM, LPR, and CV; surface analysis included the AFM, SEM, 3D profilometry, XPS, and ARXPS; theoretical testing included the MD simulations and DFT calculations that were reviewed and discussed in this chapter. Electrochemical tests are modern methods in corrosion analysis. The electrochemical kinetics of corrosion and inhibition processes are deeply investigated by the electrochemical tests. The corrosion and inhibition actions depend on the cathodic and anodic electrochemical processes on the metal interface/electrolyte solution, as the metal materials are destroyed on the anodic sites. In comparison, the hydrogen evolution and some cathodic processes have occurred on the cathodic sites. At present, PDP, EIS, EN, EFM, LPR, and CV are the most used and effective electrochemical testing methods in corrosion analysis. Surface analysis is the next important step in investigating the surface phenomenon of metal samples after corrosion and inhibition. The corrosion destruction and inhibition protection of metal surface is illustrated by surface analysis. The surface morphology on the metal surface is investigated using the AFM, SEM, 3D profilometry, XPS, and ARXPS methods. It is found that these methods accurately identify the corrosion and inhibition properties of the selected system. MD simulations and DFT calculations are correlated with the experimental results. In future research in corrosion inhibition, the PDP, EIS, EN, EFM, LPR, CV, AFM, SEM, 3D profilometry, XPS and ARXPS, MD, simulations and DFT calculations will become dominant. This is because these methods accurately describe the corrosion and inhibition processes. The corrosion inhibition performance of organic compounds can be theoretically estimated using MD simulations and DFT calculations in future research.

References

[1]   Tan J, Guo L, Yang H, Zhang F, Bakri Y. Synergistic effect of potassium iodide and sodium dodecyl sulfonate on the corrosion inhibition of carbon steel in HCl medium: A combined experimental and theoretical investigation. RSC Advances, 2020; 10: 15163–15170.
[2]   Rbaa M, Hichar A, Bazdi O, Lakhrissi Y, Ounine K, Lakhrissi B. Synthesis, characterization, and in vitro antimicrobial investigation of novel pyran derivatives based on 8–hydroxyquinoline. Beni-Suef University Journal of Basic and Applied Sciences, 2019; 12: 1–7.
[3]   Berdimurodov E, Akbarov Kh, Kholikov A. Electrochemical frequency modulation and reactivation investigation of thio glycolurils in strong acid medium. Advanced Materials Research, 2019; 1154: 122–128.
[4]   Guo L, Zhang R, Tan B, Li W, Liu H, Wu S. Locust Bean Gum as a green and novel corrosion inhibitor for Q235 steel in 0.5 M H2SO4 medium. Journal of Molecular Liquids, 2020; 310: 113239.

[5]   Dagdag O, Harfi A, Gouri M, Safi Z, Ramzi T, Jalgham T, Wazzan N, Verma Ch, Ebenso EE, Pramod Kumar U. Anticorrosive properties of Hexa (3–methoxy propan–1,2–diol) cyclotri–phosphazene compound for carbon steel in 3% NaCl medium: Gravimetric, electrochemical, DFT and Monte Carlo simulation studies. Heliyon, 2019; 5: e01340.

[6]   Rbaa M, Bazdi O, Hichar A, Lakhrissi Y, Ounine K, Lakhrissi B. Synthesis, characterization and biological activity of new pyran derivatives of 8–hydroxyquinoline. Eurasian Journal of Analytical Chemistry, 2018; 13: 19–30.

[7]   Berdimurodov E, Wang J, Kholikov A, Akbarov Kh, Burikhonov B, Umirov N. Investigation of a new corrosion inhibitor cucurbiturils for mild steel in 10% acidic medium. Advanced Engineering Forum, 2016; 18: 21–38.

[8]   Guo L, Tan B, Zuo X, Li W, Leng S, Zheng X. Eco–friendly food spice 2–Furfurylthio–3–methyl pyrazine as an excellent inhibitor for copper corrosion in sulfuric acid medium. Journal of Molecular Liquids, 2020; 317: 113915.

[9]   Rbaa M, Dohare P, Berisha A, Dagdag O, Lakhrissi L, Galai M, Zarrouk A. New Epoxy sugar based glucose derivatives as ecofriendly corrosion inhibitors for the carbon steel in 1.0 M HCl: Experimental and theoretical investigations. Journal of Alloys and Compounds, 2020; 833: 154949.

[10]   Rbaa M, Hichar A, Dohare P, Anouar El H, Lakhrissi Y, Lakhrissi B, Berredjem M, Almalki F, Rastija V, Rajabi M, Ben Hadda T, Zarrouk A. Synthesis, characterization, bio computational modeling and antibacterial study of novel pyran based on 8–hydroxyquinoline. Arabian Journal for Science and Engineering, 2021; 46: 5533–5542.

[11]   Berdimurodov E, Kholikov A, Akbarov Kh, Nakhatov I, Jurakulova N, Umirov N. Adsorption isotherm and SEM investigating of cucurbit [n]urils based corrosion inhibitors with gossypol for mild steel in alkaline media containing chloride ions. Advanced Engineering Forum, 2017; 23: 13–20.

[12]   Guo L, Bakri Y, Yu R, Tan j, Essassi M. Newly synthesized triazolopyrimidine derivative as an inhibitor for mild steel corrosion in HCl medium: An experimental and in silico study. Journal of Materials Research and Technology 2020; 9(3): 6568–6578.

[13]   Rbaa M, Benhiba F, Galai M, Abousalem AS, Ouakki M, Lai CH, Zarrouk A. Synthesis and characterization of novel Cu (II) and Zn (II) complexes of 5–{[(2–Hydroxyethyl) sulfanyl] methyl}–8–hydroxyquinoline as effective acid corrosion inhibitor by experimental and computational testings. Chemical Physics Letters, 2020; 754: 137771.

[14]   Berdimurodov E, Kholikov A, Akbarov Kh, Nuriddinova D. Polarization resistance parameters of anti–corrosion inhibitor of cucurbit [N] urils and thio glycolurils in aggressive mediums. Advanced Engineering Forum, 2018; 26: 74–86.

[15]   Guo L, Obot IB, Zheng X, Shen X, Qiang Y, Kaya S, Kaya C. Theoretical insight into an empirical rule about organic corrosion inhibitors containing nitrogen, oxygen, and sulfur atoms. Applied Surface Science, 2017; 406: 301–306.

[16]   Rbaa M, Lakhrissi B. Novel oxazole and imidazole based on 8–hydroxyquinoline as a corrosion inhibition of mild steel in HCl solution: Insights from experimental and computational studies. Surfaces and Interfaces, 2019; 15: 43–59.

[17]   Rbaa M, Galai M, Abousalem AS, Lakhrissi B, Touhami ME, Warad I, Zarrouk A. Synthetic, spectroscopic characterization, empirical and theoretical investigations on the corrosion inhibition characteristics of mild steel in molar hydrochloric acid by three novel 8–hydroxyquinoline derivatives. Ionics 2020; 13(8): 1–20.

[18]   Berdimurodov E, Kholikov A, Akbarov Kh, Xu G, Abdullah AM, Hosseini M. New anti–corrosion inhibitor (3ar,6ar)–3a,6a–di–ptolyltetrahydroimidazo[4,5–d]imidazole–2,5(1 h,3h)–dithione for carbon steel in 1 M HCl medium: Gravimetric, electrochemical, surface and quantum chemical analyses. Arabian. Journal of Chemistry, 2020; 13: 7504–7523.

[19] Hsissou R, Dagdag O, Berradi M, Bouchti El M, Assouag M, Elharfi A. Development rheological and anti–corrosion property of epoxy polymer and its composite. Heliyon, 2019; 5: e02789.

[20] Berdimurodov E, Kholikov A, Akbarov Kh, Guo L. Inhibition properties of 4,5–dihydroxy–4,5–di–p–tolylimidazolidine–2–thione for use on carbon steel in an aggressive alkaline medium with chloride ions: Thermodynamic, electrochemical, surface and theoretical analyses. Journal of Molecular Liquids, 2021; 327: 114813.

[21] Prasad D, Dagdag O, Safi Z, Wazzan N, Guo L. Cinnamoum tamala leaves extract highly efficient corrosion bio-inhibitor for low carbon steel: Applying computational and experimental studies. Journal of Molecular Liquids, 2020; 347: 118218.

[22] Hsissou R, Dagdag O, Abbout S, Benhiba F, Berradi M, El Bouchti M, Berisha A, Najat Hajjaji N, Elharfi A. Novel derivative epoxy resin TGETET as a corrosion inhibition of E24 carbon steel in 1.0 M HCl solution. Experimental and computational (DFT and MD simulations) methods. Journal of Molecular Liquids, 2019; 284: 182–192.

[23] Rbaa M, Fardioui M, Verma C, Abousalem AS, Galai M, Ebenso EE, Zarrouk A. 8–Hydroxyquinoline based chitosan derived carbohydrate polymer as biodegradable and sustainable acid corrosion inhibitor for mild steel: Experimental and computational analyses. International Journal of Biological Macromolecules, 2020; 155: 645–655.

[24] Berdimurodov E, Kholikov A, Akbarov Kh, Guo L, Abdullah AM, Elik M. A gossypol derivative as an efficient corrosion inhibitor for St2 steel in 1 M HCl + 1 M KCl: An experimental and theoretical investigation. Journal of Molecular Liquids, 2021; 328: 115475.

[25] Hsissou R, Benhiba F, Dagdag O, El Bouchti M, Nouneh K, Assouag M, Briche S, Zarrouk A, Elharfi A. Development and potential performance of prepolymer in corrosion inhibition for carbon steel in 1.0 M HCl: Outlooks from experimental and computational investigations. Journal of Colloid and Interface Science, 2020; 574: 43–60.

[26] Rbaa M, Benhiba F, Hssisou R, Lakhrissi Y, Lakhrissi B, Touhami ME, Zarrouk A. Green synthesis of novel carbohydrate polymer chitosan oligosaccharide grafted on d–glucose derivative as bio–based corrosion inhibitor. Journal of Molecular Liquids, 2021; 322: 114549.

[27] Berdimurodov E, Kholikov A, Akbarov Kh, Obot IB, Guo L. Thioglycoluril derivative as a new and effective corrosion inhibitor for low carbon steel in a 1 M HCl medium: Experimental and theoretical investigation. Journal of Molecular Structure, 2021; 1234: 130165.

[28] Haldhar R, Prasad D, Bahadur I, Dagdag O, Berisha A. Evaluation of Gloriosa superba seeds extract as corrosion inhibition for low carbon steel in sulfuric acidic medium: A combined experimental and computational studies. Journal of Molecular Liquids, 2021; 323: 114958.

[29] Berdimurodov E, Kholikov A, Akbarov Kh, Guo L. Experimental and theoretical assessment of new and eco–friendly thioglycoluril derivative as an effective corrosion inhibitor of St2 steel in the aggressive hydrochloric acid with sulfate ions. Journal of Molecular Liquids, 2021; 335: 116168.

[30] Kumar D, Kazi M, Alqahtani MS, Syed R, Berdimurodov E. N – Hydroxybenzothioamide derivatives as green and efficient corrosion inhibitors for mild steel: Experimental, DFT and MC simulation approach. Journal of Molecular Structure, 2021; 1241: 130648.

[31] Shahmoradi AR, Ranjbarghanei M, Javidparvar AA, Guo L, Berdimurodov E, Ramezanzadeh B. Theoretical (atomic–DFT&molecular–MD), surface/electrochemical investigations of walnut fruit green husk extract as effective–biodegradable corrosion mitigating materials of a steel electrode in 1M HCl electrolyte. Journal of Molecular Liquids, 2021; 338: 116550.

[32] Berdimurodov E, Kholikov A, Akbarov Kh, Guo L, Kaya S, Verma DK, Rbaa M, Dagdag O. New and green corrosion inhibitor based on new imidazole derivate for carbon steel in 1 M HCl medium: Experimental and theoretical analyses. International Journal of Engineering Research in Africa, 2022; 58: 11–44.

[33] Verma DK, Kazi M, Alqahtani MS, Syed R, Berdimurodov E, Kaya S, Salim R, Asatkar A, Haldhar R. N–hydroxybenzothioamide derivatives as green and efficient corrosion inhibitors for mild steel: Experimental, DFT and MC simulation approach. Journal of Molecular Structure, 2021; 1241: 130648.

[34] Shahmoradi AR, Ranjbarghanei M, Javidparvar AA, Guo L, Berdimurodov E, Ramezanzadeh B. Theoretical and surface/electrochemical investigations of walnut fruit green husk extract as effective inhibitor for mild–steel corrosion in 1M HCl electrolyte. Journal of Molecular Liquids, 2021; 338: 116550.

[35] Dagdag O, Hsissou R, Safi Z, Haldhar R, Berdimurodov E, Bouchti ME, Wazzan N, Hamed O, Jodeh S, Gouri ME. Rheological and simulation for macromolecular matrix epoxy bi–functional aromatic amines. Polymer Bulletin, 2021; 27: 1–7.

[36] Bahgat Radwan A, Mannah CA, Sliem MH, Al–Qahtani NH, Okonkwo PC, Berdimurodov E, Mohamed AM, Abdullah AM. Electrospun highly corrosion–resistant polystyrene–nickel oxide super hydrophobic nanocomposite coating. Journal of Applied Electrochemistry, 2021; 6: 1–4.

[37] Rbaa M, Oubihi A, Hajji H, Tüzün B, Hichar A, Berdimurodov E, Ajana MA, Zarrouk A, Lakhrissi B. Synthesis, bioinformatics and biological evaluation of novel pyridine based on 8–hydroxyquinoline derivatives as antibacterial agents: DFT, molecular docking and ADME/T studies. Journal of Molecular Structure, 2021; 1244: 130934.

[38] Eid AM, Shaaban S, Shalabi K. Tetrazole-based organoselenium bi-functionalized corrosion inhibitors during oil well acidizing: Experimental, computational studies, and SRB bioassay. Journal of Molecular Liquids, 2020; 298: 111980.

[39] Obot IB, Meroufel A, Onyeachu IB, Alenazi A, Sorour AA. Corrosion inhibitors for acid cleaning of desalination heat exchangers: Progress, challenges and future perspectives. Journal of Molecular Liquids, 2019; 296: 111760.

[40] Singh A, Ituen EB, Ansari KR, Chauhan DS, Quraishi MA. Surface protection of X80 steel using Epimedium extract and its iodide–modified composites in simulated acid wash solution: A greener approach towards corrosion inhibition. New Journal of Chemistry 2019; 43(22): 8527–8538.

[41] Sliem MH, Afifi M, Radwan AB, Fayyad EM, Shibl MF, Heakal FE–T, Abdullah AM. AEO7 surfactant as an eco–friendly corrosion inhibitor for carbon steel in HCl solution. Scientific Reports 2019; 9(1): 1–16.

[42] Madkour LH, Kaya S, Guo L, Kaya C. Quantum chemical calculations, molecular dynamic (MD) simulations and experimental studies of using some azo dyes as corrosion inhibitors for iron. Part 2: Bis–azo dye derivatives. Journal of Molecular Structure, 2018; 163: 397–417.

[43] Solomon MM, Gerengi H, Umoren SA, Essien NB, Essien UB, Kaya E. Gum Arabic-silver nanoparticles composite as a green anticorrosive formulation for steel corrosion in strong acid media. Carbohydrate Polymers, 2018; 181: 43–55.

[44] Kousar K, Ljungdahl T, Wetzel A, Dowhyj M, Oskarsson H, Walton AS, Walczak MS, Lindsay R. An exemplar imidazoline surfactant for corrosion inhibitor studies: Synthesis, characterization, and physicochemical properties. Journal of Surfactants and Detergents, 2020; 23: 225–234.

[45] Obot IB, Onyeachu IB, Umoren SA. Alternative corrosion inhibitor formulation for carbon steel in CO2–saturated brine solution under high turbulent flow condition for use in oil and gas transportation pipelines. Corrosion Science, 2019; 159: 108140.

[46] Onyeachu IB, Obot IB, Sorour AA, Abdul–Rashid MI. Green corrosion inhibitor for oilfield application I: Electrochemical assessment of 2–(2–pyridyl)benzimidazole for API X60 steel under sweet environment in NACE brine ID196. Corrosion Science, 2019; 150: 183–193.

[47] Qiang Y, Guo L, Li H, Lan X. Fabrication of environmentally friendly Losartan potassium film for corrosion inhibition of mild steel in HCl medium. Chemical Engineering Journal, 2020; 406: 126863.

[48] Singh A, Ansari K, Chauhan DS, Quraishi M, Kaya S. Anti–corrosion investigation of pyrimidine derivatives as green and sustainable corrosion inhibitor for N80 steel in highly corrosive environment: Experimental and AFM/XPS study. Sustainable Chemistry and Pharmacy, 2020; 16: 100257.

[49] Mohammadkhani R, Ramezanzadeh M, Akbarzadeh S, Bahlakeh G, Ramezanzadeh B. Graphene oxide nanoplatforms reduction by green plant sourced organic compounds for construction of an active anti-corrosion coating: Experimental/electronic–scale DFT–D modeling studies. Chemical Engineering Journal, 2020; 397: 125433.

[50] Ansari K, Chauhan DS, Quraishi M, Mazumder MA, Singh A. Chitosan Schiff base: An environmentally benign biological macromolecule as a new corrosion inhibitor for oil & gas industries. International Journal of Biological Macromolecules, 2020; 144: 305–315.

[51] Ismail MC, Yahya S, Raja PB. Antagonistic effect and performance of CO2 corrosion inhibitors: Water chemistry and ionic response. Journal of Molecular Liquids, 2019; 293: 111504.

[52] Onyeachu IB, Quraishi MA, Obot IB, Haque J. Newly synthesized pyrimidine compound as CO2 corrosion inhibitor for steel in highly aggressive simulated oilfield brine. Journal of Adhesion Science and Technology, 2019; 33: 1226–1247.

[53] El–Haddad MAM, Radwan AB, Sliem MH, Hassan WMI, Abdullah AM. Highly efficient eco-friendly corrosion inhibitor for mild steel in 5 M HCl at elevated temperatures: Experimental & molecular dynamics study. Scientific Reports 2019; 9(1): 1–15.

[54] Eduok U, Ohaeri E, Szpunar J. Conversion of imidazole to N–(3–Aminopropyl)imidazole toward enhanced corrosion protection of steel in combination with carboxymethyl chitosan grafted poly (2–methyl–1–vinylimidazole). Industrial & Engineering Chemistry Research 2019; 58(17): 7179–7192.

[55] Corrales–Luna M, Le Manh T, Romero–Romo M, Palomar–Pardavé M, ArceEstrada EM. 1–ethyl 3–methylimidazolium thiocyanate ionic liquid as corrosion inhibitor of API 5L X52 steel in H2SO4 and HCl media. Corrosion Science, 2019; 153: 85–99.

[56] Luna MC, Le Manh T, Sierra RC, Flores JVM, Rojas LL, Estrada EMA. Study of corrosion behavior of API 5L X52 steel in sulfuric acid in the presence of ionic liquid 1–ethyl 3–methylimidazolium thiocyanate as corrosion inhibitor. Journal of Molecular Liquids, 2019; 289: 111106.

[57] Raj R, Morozov Y, Calado LM, Taryba MG, Kahraman R, Shakoor A, Montemor MF. Inhibitor loaded calcium carbonate microparticles for corrosion protection of epoxy-coated carbon steel. Electrochimica Acta, 2019; 319: 801–812.

[58] Morozov Y, Calado LM, Shakoor RA, Raj R, Kahraman R, Taryba MG, Montemor MF. Epoxy coatings modified with a new cerium phosphate inhibitor for smart corrosion protection of steel. Corrosion Science, 2019; 159: 108128.

[59] Berdimurodov E. Adsorption equilibrium, kinetics, thermodynamics and dynamic separation of magnesium and calcium ions from industrial wastewater by new strong acid cation resin of SPVC. Pakistan Journal of Analytical & Environmental Chemistry, 2021; 22: 127–138.

[60] Dagdag O, Hsissou R, Safi Z, Haldhar R, Berdimurodov E, Bouchti ME El, Wazzan N, Hamed O, Jodeh S, Gouri M. Rheological and simulation for macromolecular matrix epoxy bi-functional aromatic amines. Polymer Bulletin, 2021; 1: 1–17.

[61] Yarkulov A, Umarov B, Rakhmatkarieva F, Kattaev N, Akbarov Kh, Berdimurodov E. Diacetate cellulose-silicon bio nanocomposite adsorbent for recovery of heavy metal ions and benzene vapors: An experimental and theoretical investigation. Bio interface Research in Applied Chemistry, 2022; 12: 2862–2880.

[62] Berdimurodov E, Kholikov A, Akbarov Kh, Guo L. Inhibition properties of 4,5-dihydroxy-4,5-di-p-tolylimidazolidine-2-thione for use on carbon steel in an aggressive alkaline medium with chloride ions: Thermodynamic, electrochemical, surface and theoretical analyses. Journal of Molecular Liquids, 2021; 327: 114813.

[63] Rbaa M, Oubihi A, Hajji H, Tüzün B, Hichar A, Anouar EH, Berdimurodov E, Ajana MA, Zarrouk A, Lakhrissi B. Synthesis, bioinformatics and biological evaluation of novel pyridine based on 8-hydroxyquinoline derivatives as antibacterial agents: DFT, molecular docking and ADME/T studies. Journal of Molecular Structure, 2021; 1244: 130934.

[64] Berdimurodov E, Kholikov A, Akbarov Kh, Guo L. Experimental and theoretical assessment of new and eco-friendly thioglycoluril derivative as an effective corrosion inhibitor of St2 steel in the aggressive hydrochloric acid with sulfate ions. Journal of Molecular Liquids, 2021; 335: 116168.

[65] Berdimurodov E, Kholikov A, Akbarov Kh, Guo L, Kaya S, Katin KP, Verma DK, Rbaa M, Dagdag O, Haldhar R. Novel gossypol–indole modification as a green corrosion inhibitor for low–carbon steel in aggressive alkaline–saline solution. Colloids and Surfaces. A, Physicochemical and Engineering Aspects, 2022; 637: 128207.

[66] Berdimurodov E, Kholikov A, Akbarov Kh, Guo L, Kaya S, Kumar Verma D, Rbaa M, Dagdag O. Novel glycoluril pharmaceutically active compound as a green corrosion inhibitor for the oil and gas industry. Journal of Electroanalytical Chemistry, 2022; 907: 116055.

[67] Haldhar R, Kim S-C, Berdimurodov E, Verma DK, Hussain CM. Corrosion inhibitors: Industrial applications and commercialization. *Sustainable Corrosion Inhibitors II: Synthesis, Design, and Practical Applications*. American Chemical Society (ACS), 2021, 10–219.

[68] Dagdag O, Berisha A, Mehmeti V, Haldhar R, Berdimurodov E, Hamed O, Jodeh S, Lgaz H, Sherif E-SM, Ebenso EE. Epoxy coating as effective anti-corrosive polymeric material for aluminum alloys: Formulation, electrochemical and computational approaches. Journal of Molecular Liquids, 2021; 346: 117886.

[69] Berdimurodov E, Kholikov A, Akbarov Kh, Guo L, Kaya S, Katin KP, Verma DK, Rbaa M, Dagdag O. Novel cucurbit[6]uril-based [3]rotaxane supramolecular ionic liquid as a green and excellent corrosion inhibitor for the chemical industry. Colloids and Surfaces. A, Physicochemical and Engineering Aspects, 2022; 633: 127837.

[70] Berdimurodov E, Guo L, Kholikov A, Akbarov Kh, Zhu M. MOFs-based corrosion inhibitors. *Supramolecular Chemistry in Corrosion and Biofouling Protection*. Taylor and Francis, CRC Press, 2021, 287–305.

[71] Bahgat Radwan A, Mannah CA, Sliem MH, Al-Qahtani NHS, Okonkwo PC, Berdimurodov E, Mohamed AM, Abdullah AM. Electrospun highly corrosion-resistant polystyrene-nickel oxide superhydrophobic nanocomposite coating. Journal of Applied Electrochemistry, 2021; 51: 1605–1618.

[72] Zhu M, Guo L, He Z, Marzouki R, Zhang R, Berdimurodov E. Insights into the newly synthesized N-doped carbon dots for Q235 steel corrosion retardation in acidizing media: A detailed multidimensional study. Journal of Colloid and Interface Science, 2022; 608: 2039–2049.

[73] Berdimurodov E, Kholikov A, Akbarov Kh, Guo L, Kaya S, Katin KP, Kumar VD, Rbaa M, Dagdag O, Haldhar R. Novel bromide-cucurbit[7]uril supramolecular ionic liquid as a green corrosion inhibitor for the oil and gas industry. Journal of Electroanalytical Chemistry, 2021; 901: 115794.

[74] Berdimurodov E, Verma DK, Kholikov A, Akbarov Kh, Guo L. The recent development of carbon dots as powerful green corrosion inhibitors: A prospective review. Journal of Molecular Liquids, 2021; 351: 118124.

[75] Verma DK, Kazi M, Alqahtani MS, Syed R, Berdimurodov E, Kaya S, Salim R, Asatkar A, Haldhar R. N–hydroxybenzothioamide derivatives as green and efficient corrosion inhibitors for mild steel: Experimental, DFT and MC simulation approach. Journal of Molecular Structure, 2021; 1241: 130648.

Rajimol Puthenpurackal Ravi, Sarah Bill Ulaeto,
Thazhavilai Ponnu Deva Rajan, Kokkuvayil Vasu Radhakrishnan

8 Natural product-based multifunctional corrosion inhibitors for smart coatings

Abstract: Responsive or smart materials are those that can dynamically adapt their properties to an external stimulus. Temperature, light, pressure, wettability, pH variation, harsh corrosive species, or other micro-environmental changes can stimulate smart coatings. Because of their intrinsic features and chemistry on modification, smart coatings have earned recognition in all areas of sciences. Examples of smart coatings include corrosion-sensing, antimicrobial, antifouling, self-healing, fire-retardant, self-cleaning, and super-hydrophobic systems. Even though many intelligent coatings are developed with synthetic organic and inorganic materials, bio-based equivalents are effective competitors as they are cheap, environmentally friendly, and are easily renewable. By the incorporation of bio-based components in coatings, anticorrosion systems with other smart properties have become well-established. The phytochemicals containing heteroatoms can donate the lone pair, preventing metal corrosion. The higher carbon content in biomass helps in fire-retardant coatings. Many plant extracts and phytochemicals have antimicrobial properties, which contribute immensely to developing antimicrobial, antifouling, and self-cleaning coatings. In the event of new epidemics and pandemics, corrosion resistant coatings with antimicrobial and self-cleaning properties are very relevant. The inherent properties of biomasses and individual phytochemicals are responsible for the smart properties of these coatings. Vegetable oil-based epoxy coatings are excellent replacements for petroleum based epoxy products that have numerous side effects. Smart coatings help in reducing manpower consumption and improve energy efficiency. There are many biomass-derived smart coating products available in the market. But among these coatings, only anticorrosion coatings with self-healing and self-cleaning coatings have been thoroughly researched. Bio-based smart coatings are hot areas of research with a lot of scope and unexplored possibilities.

Keywords: Anticorrosion, smart coatings, biomass, phytochemicals, self-cleaning, self-healing, antimicrobial, antifouling, super-hydrophobic, anti-graffiti, anti-icing, environmental friendly

Acknowledgments: Rajimol P. R. acknowledges UGC, India, for research fellowship. The authors are thankful to the Director, CSIR-National Institute of Interdisciplinary Science and Technology (NIIST), Trivandrum, for providing the environment to carry out the research and preparation of this chapter.

https://doi.org/10.1515/9783110760583-008

8.1 Introduction

The gradual destruction of pure metals and their alloys into their more stable forms, such as oxides, hydroxides, carbonates, or sulfides due to environmental or electrochemical factors, lead to corrosion. It is an unwanted but spontaneous process, resulting in many disasters and has a substantial economic impact that affects the development of a nation. Many infrastructure failures, imperfections, accidents, and financial losses occur due to this unwanted phenomenon. On an average, a developed country loses about 4% of its gross domestic product (GDP) every year due to corrosion. The resultant effect can be experienced in growth retardation in the nation's diverse fields [1].

Application of various corrosion inhibitors, cathodic protection, anodic protection, and protective coatings are the standard methods used to prevent or reduce corrosion rates in common metals and alloys. The selection of a particular method to be used will be based on the metal/alloy used, environmental conditions, expenses involved, corrosive media, the areas of application of the metal surface, and so on. The application of corrosion inhibitors is the most convenient and commonly used method. A small amount of corrosion inhibitor can provide a very fair rate of corrosion protection. Several organic, inorganic, and heavy metal inhibitors are readily available at a low cost.

8.1.1 Natural product-based corrosion inhibitors

Previously, synthetic organic and inorganic inhibitors that were used were of meagre cost and offered good inhibition efficiency. Chromates, phosphates, heavy metals, and azole derivatives were the widely used and most effective corrosion inhibitors, but now they are banned in many countries. Chromate is a very effective inhibitor for all types of metallic corrosion because of its high oxidation property. It acts as a mixed inhibitor and reduces the overall electrochemical activity by inhibiting both anodic and cathodic reactions. Minute concentrations of the ion can inhibit corrosion to a great extent. The loading capacity of the chromates in coatings is very low; thus, the chromates tend to leach out of the surface, which in turn lowers the efficiency, at times [2]. Despite the higher rate of metal protection, the chromates' leachability and higher oxidation properties make them environmentally unfriendly. Direct interaction with the skin, inhalation, and ingestion through polluted water, lead to chromium pollution in the human body. The hexavalent chromium ions are carcinogenic and can cause DNA and RNA damage [3]. This challenge due to the use of hexavalent chromium has encouraged the exploration of natural products as corrosion inhibitors.

Plant-based compounds and extracts have been researched over time and found to be effective corrosion inhibitors. Plant biomass is an environmentally friendly, low-cost renewable source. The antibacterial, antiviral, and antifungal capabilities of plant-based compounds reflect their multifunctional properties. These characteristics have made them valuable enough to be considered for various intelligent coatings. Many plant oil derivatives are used as monomers for making eco-friendly corrosion-resistant polymers. Epoxidized castor oil, epoxidized soy-bean oil, and cardanol derivatives are excellent choices. The addition of a strong inhibitor to this polymer matrix can increase the ability manifold. Phytochemicals are excellent choices for this application.

The whole plant or any particular part of the plant, such as the leaf, seed, root, and stem is extracted with a specific solvent. The isolated phytochemicals can act as corrosion inhibitors. The activity of the extract depends on the nature of the major phytochemicals present in it. Phytochemicals can be classified into primary and secondary metabolites, depending on their biological function. Primary metabolites are responsible for the growth, and secondary metabolites are responsible for the defense mechanism of plants against insects, other parasites, and disease-causing microbes. For quantitative analysis of phytochemicals, many chromatographic techniques, such as column chromatography, liquid chromatography, high-performance liquid chromatography (HPLC), and gas chromatography (GC), can be applied. Maceration, percolation, and soxhlet extraction methods are prominently employed in phytochemical screening studies, but there are some advanced methods, such as ultrasound-assisted extraction (UAE), supercritical fluid extraction (SFE), microwave-assisted extraction (MAE), and accelerated solvent extraction [4].

Extraction can be broadly classified as isolation with organic solvents and green extraction. In extraction with organic solvents of varying polarity, dried materials are soaked in and stirred with different organic solvents, such as hexane, ethyl acetate, chloroform, acetone, DCM, ethanol, or methanol. The active phytochemicals dissolve in the respective solvents, depending on their polarity or the functional groups present. After repeated extraction with these solvents, the solvents are removed under reduced pressure and optimum temperature. The solid residue obtained is vacuum dried and kept in a refrigerator, and column chromatography is the main technique used for the isolation of secondary metabolites.

Generally, phytochemicals containing heteroatoms or pi-electron clouds are found to be more effective in corrosion inhibition. Alkaloid, carboxylic acids, and its derivatives, such as esters and lactones, ketones, terpenes, and polyphenols such as flavanoids, flavanols, chalcones, flavones, coumarins, tannins, flavanones, lignans, stilbenes, and organosulfur compounds are potent candidates for corrosion inhibition studies. The structures of some of the compounds from these categories are listed below. The lone pair of electrons available on the heteroatoms, such as nitrogen, oxygen, phosphorus, and sulfur, can interact with the metal ion, and thus stabilize and prevent them from further redox reactions. In many phytochemicals, there is an aromatic system or pi-electron cloud. These loosely bonded electrons can interact with the metal atoms and ions by electron coordination. Selected examples of phytochemical compounds with the expected corrosion inhibiting properties are shown in Figure 8.1.

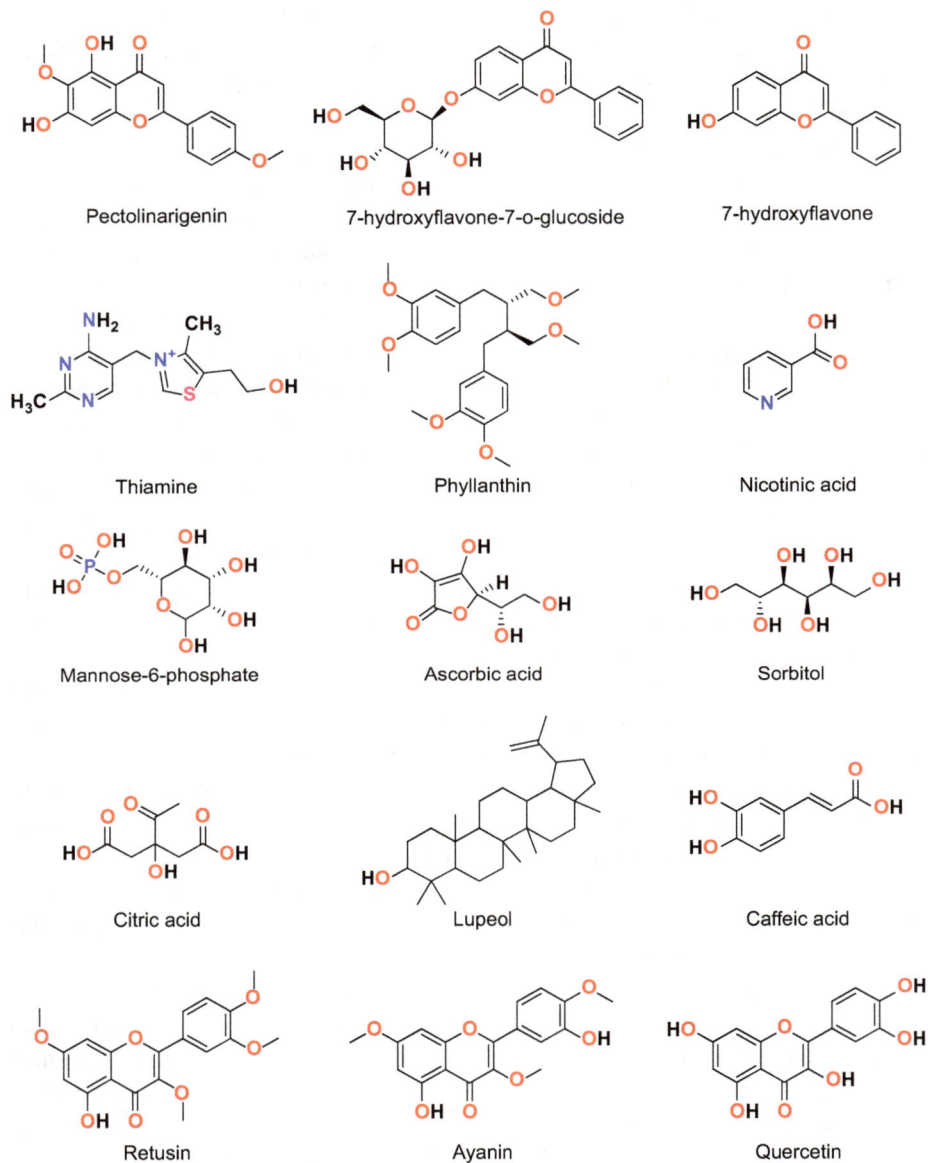

Figure 8.1: Examples of natural product compounds with the expected corrosion inhibition property.

8.2 Smart coatings

Coatings have traditionally been thought of as a passive layer that is insensitive to the environment. Responsive or smart materials are those that can dynamically adapt their properties to an external stimulus. Temperature, light, pressure, wettability, pH variation, temperature, harsh corrosive species, or other microenvironmental changes can stimulate smart coatings. "Stimulus-responsive," "intelligent," or "environmentally sensitive" films are other terms for smart coatings. Because of their intrinsic features and chemistry on modification, smart coatings have earned recognition in all areas of sciences. Examples of smart coatings include corrosion-sensing, antimicrobial, antifouling, self-healing, conductive, and superhydrophobic systems. They are made with both passive and active species; when employed in an application, the uses of their rapid response, based on the appropriate stimuli, can be realized. Innovative coatings are creatively developed for various applications and their value lies in their capacity to react to numerous cycles over a long period. By the incorporation of micro/nanoparticle and combining organic and inorganic groups, intelligent coatings have achieved outstanding qualities than conventional coatings [5, 6].

Stimuli-responsive coatings are employed in various sectors and for extensive purposes. These include medical, aviation, textile, transportation, construction, electronics, military, etc. to provide various services, including corrosion protection. Without physical assistance, they are seen as pragmatic candidates. In many industrial applications, the purpose of smart coatings is to improve system efficiency by decreasing inspection times, lowering repair costs, and reducing instrument downtime. It is a class of coating that has the potential to have a significant impact on society [7].

Self-healing coatings can mend themselves, stopping the further crack spread and, in some cases, a complete cure of the crack zone. Corrosion-sensing and pressure-sensing coatings are active sensing coatings that notify promptly when a change in the microenvironment occurs. Optically active materials are used to create optically active coatings, which are relevant for smart window coatings. Such windows limit incoming IR radiation from the sun throughout the hot season and retain temperature within the apartment during the cold weather, improving energy efficiency. Anti-graffiti coatings and self-cleaning coatings are examples of easy-to-clean coatings. Bioactive coatings include antifouling and antimicrobial coatings. Non-intumescent and intumescent coatings both have the potential to retard fire. Smart coatings also include antifingerprint, antireflective, anti-icing, and antifogging coatings, to mention a few [5]. Smart coatings are a boon to humanity since they reduce wastage of human resources, maintenance costs, disease propagation, and avert unanticipated mishaps.

8.2.1 Properties of natural products beneficial for smart coatings

Natural products are low-cost renewable components. The phytochemicals containing heteroatoms such as nitrogen, phosphorus, oxygen, and sulfur can donate the lone pair, preventing metal corrosion. Corrosion inhibitors are often substances with varying oxidation states or can donate electrons to the metal matrix. The electron cloud is found in unsaturated and aromatic systems, and it is one of the extra benefits of phytochemicals as corrosion inhibitors. Furthermore, bio-based products are undeniably high in carbon. In fire-retardant coatings, an enormous amount of carbon is required [8]. Many plant extracts and phytochemicals have antimicrobial properties, which contribute immensely to developing antimicrobial, antifouling, and self-cleaning coatings.

8.3 Natural product-based smart coatings

Even though many intelligent coatings are developed with synthetic organic and inorganic materials, bio-based equivalents are effective competitors. With the substitution of synthetic components by bio-based components, self-cleaning/healing, anticorrosion coatings have become well-established. Identifying and isolating natural substances and their synthetic derivatives with this multifunctional feature should prioritize future research.

8.3.1 Anticorrosion coatings

Organic coatings create a barrier against the diffusion of oxygen, water, and other harsh ions to the metal surface, thus reducing the rate of corrosion. Organic solvent-based polyurethanes were the prime candidates for corrosion inhibition in the past few decades. Nevertheless, the increasing environmental hazards due to volatile organic compound (VOC) emission, depleting petroleum sources, and health issues associated with petroleum-based epoxy led to the invention of renewable waterborne polyurethanes (WPU). The most promising renewable sources were vegetable oils, as they have relatively low prices, low toxicity, and inherent biodegradability. Reactive functional groups are naturally present in some oils, such as hydroxyl in castor oil and epoxy in vernonia oil. Free radical or cationic polymerization is enough for these vegetable oils to produce polymers without any additional chemical modification. However, oils bearing functional groups, such as esters and double bonds in triglycerides, require a further chemical modification prior to converting functional monomers into excellent polymers [9]. However, in the last few decades, many hybrid coatings were prepared by the effective incorporation of various nanoparticle fillers, such as ZnO, TiO_2, Cr_2O_3, and Fe_2O_3, to improve the corrosion inhibition efficiency of organic coatings [10].

Hegde et al. created a variety of linseed oil epoxy coatings. Two different hardeners obtained from linseed oil were used to cure epoxidized linseed oil (ELO). The hardeners were produced by reacting linseed oil with maleic anhydride to produce melanized linseed oil (H1), and the second hardener, H2, was produced by reacting H1 with diethylenetriamine in ethyl acetate. Graphene-based fillers in the nanoscale were utilized in various percentages. The percentages of graphene oxide rGO-ELO-H2 coating exhibited superior inhibition efficiency of 99.98% and excellent stability in 3.5 wt% NaCl. The Nyquist plot in Figure 8.2 reveals large diameters of capacitive loops from the graphene filler-loaded bio-based coatings, strongly indicative of corrosion inhibition. This research demonstrates a simple and environmentally friendly method for integrating nanofillers into vegetable oil-based coatings, which could be extremely valuable in the coatings and polymer industries [11].

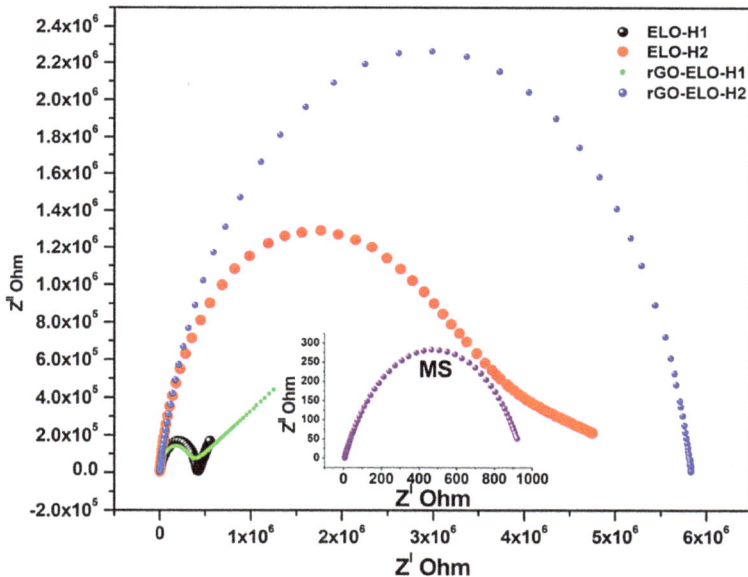

Figure 8.2: Nyquist plot for bare mild steel and coated mild steel coupons after 2 h of immersion in a saline medium (adapted with permission from reference [11]. Copyright M.B. Hegde, K.N.S. Mohana, K. Rajitha, A.M. Madhusudhana, some rights reserved; exclusive licensee Elsevier. Distributed under a Creative Commons Attribution License 4.0 (CC BY) https://creativecommons.org/licenses/by/4.0/).

In another study, *Gossypium arboreum* (cottonseed) plant oil was turned into fatty amide and then esterified with various dicarboxylic acids and anhydrides of renewable nature to produce a range of polyesteramide polyols to substitute petroleum counterparts. Sebacic acid, succinic acid, tartaric acid, maleic acid, and azelaic acid (all isolated from bio-based sources) were among the bio-based renewable dicarboxylic acids and anhydrides used in this experiment to replace petroleum counterparts. Compared to petro-based commercial coatings, vegetable oil-based solutions exhibited comparable or greater corrosion inhibition action [12]. Figure 8.3 is a pictorial representation of the corrosion resistance of the bio-based PU coatings on mild steel surface in 3.5 wt% NaCl solution, before and after immersion.

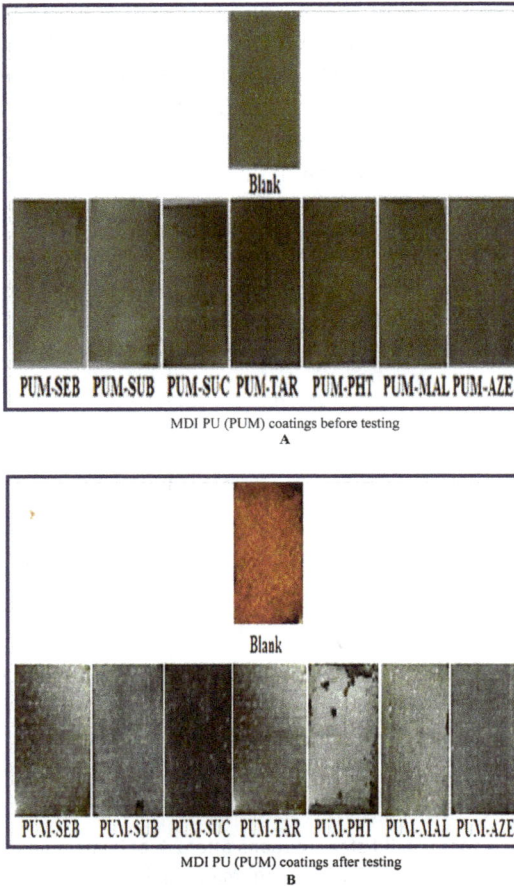

Figure 8.3: The corrosion resistance of different PU coatings on MS surface in a saline medium (A-before testing, B-after testing) (adapted with permission from reference [12]. Copyright C.K. Patil, S.D. Rajput, R.J. Marathe, R.D. Kulkarni, H. Phadnis, D. Sohn, P.P. Mahulikar, V. V Gite, some rights reserved; exclusive licensee Elsevier. Distributed under a Creative Commons Attribution License 4.0 (CC BY) https://creativecommons.org/licenses/by/4.0/).

Bin Liang et al. developed two types of renewable UV-curable active monomers, derived from tung oil: tung-acid-maleic triacrylate (TOAH) and tung-maleic tetraacrylate (TMPG). The two polymers had differing chemical, physical, and mechanical properties. Combining these two bio-based monomers, a monomer could be obtained with comparable balanced properties and the best performance in all properties. Pure TMPG addition-cured films were rigid materials with good modulus and thermal stability, whereas pure TOAH addition-cured films had the lowest strength and modulus. To improve the toughness of TMPG and increase the strength of the TOAH, copolymerization of the two active monomers with varied weight ratios was carried out. The UV-curing efficacy was evaluated using photo-DSC investigations, and thermogravimetric measurement revealed that the maximum decomposition temperature of the copolymer coatings was all above 420 °C. As a result, it had a faster curing rate, increased hardness, superior thermal stability, and corrosion inhibition [13].

Silva et al. developed a bio-based corrosion-resistant epoxy coating for steel using agricultural waste material, an outstanding waste-to-wealth management demonstration involving direct valorization of the liquid from cashew nutshells (CNSL). For the epoxidation of cardanol and cardol, a solvent-free in situ technique was adopted to get epoxidized cashew nut shell liquid (e-CNSL). The results of the different curing agents, diethylenetriamine (DETA), isophorone diamine (IPDA), and perfluorodecanoic acid (PFDA) were compared to a standard DGEBA/IPDA coating. Electrochemical impedance spectroscopy was used to assess the coatings' anticorrosive capabilities. The findings revealed that e-CNSL-based coatings had much potential as bio-based corrosion prevention coatings and they could be an excellent alternative to BPA-based products. Due to the chemical composition of CNSL, which combines a phenolic core with a lipidic side chain, the bio-based coatings demonstrated a good balance of flexibility and wettability properties. Excellent chemical resistance, thermal stability, and anticorrosion performance were also achieved, indicating that the bio-based coatings could be considered promising sustainable anticorrosive solutions for metal surfaces [14].

8.3.2 Self-healing coatings

Coatings effectively protect a metal surface from corrosion, but there are chances that this coating can get damaged during shipment, installation, or service; hence lowering their overall performance [15]. Organic coatings based on epoxy resins were one of the best choices for protecting a metal surface because they prohibit direct contact with oxygen and water. However, standard epoxy coatings fail to prevent the flaw from propagating through the metal matrix when a fracture occurs. As a result, coatings that intelligently respond to damage and avoid degradation material are in high demand [16].

Self-healing coatings belong to the class of intelligent repair coatings, inspired by the natural wound healing mechanism. The concept of self-repair is a prominent feature of the living world that has grabbed the interest of medical and pharmaceutical researchers, and it is highly relevant nowadays in the development of corrosion-resistant materials. Jud and Kausch proposed self-healing in 1979, based on molecular interdiffusion along with crack interfaces, which was further improved by White and coworkers in 2001 by incorporating repairing fluids and catalyst grains into the matrix material. These auto repair coatings have several advantages over conventional coatings, including i) automatic damage repair; ii) auto-preservation of the aesthetics of the coatings and plastic products; and iii) reestablishment of the mechanical integrity of objects that require high strength [17]. Self-healing materials are no longer a pipe dream, and we are not far from time when man-made materials can reclaim their structural integrity after a failure. Building cracks, for example, can close on their own, and scratches on automobile bodies can revert to their previous gleaming appearance [1].

Adding self-healing characteristics to man-made materials might occasionally fail to act without stimulation. As a result, self-healing can be classified into autonomic self-healing and non-autonomic self-healing. Non-autonomic self-healing systems require prompting from the outside to begin their function. While autonomous self-healing doesn't necessitate outside assistance; the damage itself initiates repair mechanisms. There are several methods for obtaining a material's self-healing capability, such as

- Release of a healing agent or catalyst
- Reversible chemical bonds like a Diels Alder reaction
- Shape memory effect
- Nanoparticle migration
- Co-deposition.

The release of healing agents from the matrix is one of the critical steps in a self-healing mechanism. Direct incorporation into the matrix is many times detrimental to the service life of the coating. Thus, inhibitor-holding fillers are used either in the macroscale or the nanoscale. Methods, such as microcapsule embedment, hollow fiber embedment, and microvascular systems, can store the healing agent or monomers. During synthesis, liquid active agents, such as monomers, catalysts, dyes, and hardeners, are immersed in polymeric systems in microparticles, empty fibers, or channels. These reservoirs are broken in the case of a fracture, and active substances are spilled into the cracks by capillary force, where they solidify with predispersed catalysts and cure the crack. The spread of the crack is the driving factor in this autonomic repair [1]. Major self-healing mechanisms in polymers are described in Figure 8.4. The illustration demonstrates how curing species in micro/nanocapsules that are incorporated in polymer matrices cause auto-repairing [18].

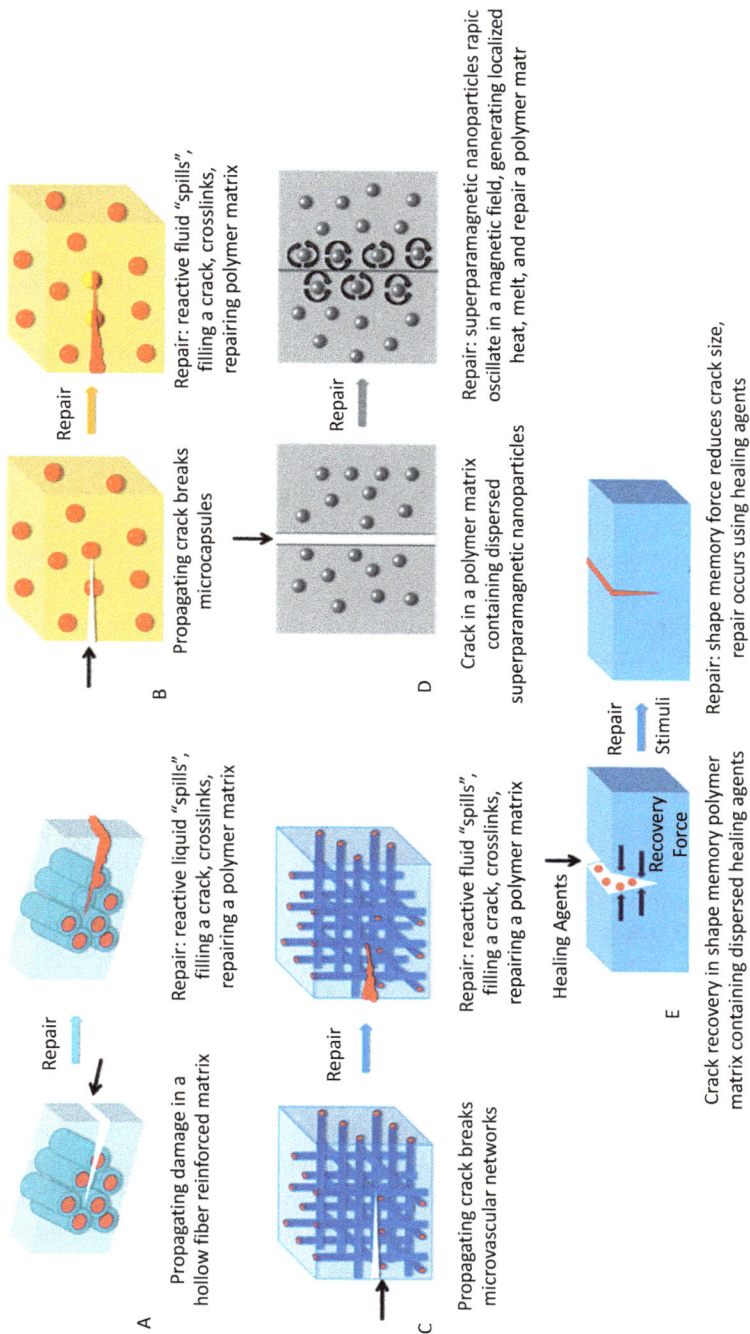

Figure 8.4: (A–E) Self-repairing mechanisms in polymer matrices with healing agents in nanocontainers (adapted with permission from reference [18]. Copyright Y. Yang, M.W. Urban, some rights reserved; exclusive licensee Royal Society of Chemistry. Distributed under a Creative Commons Attribution License 4.0 (CC BY) https://creativecommons.org/licenses/by/4.0/).

Many initiatives have been made to replace fossil feedstock chemicals with biomass-derived compounds, mainly vegetable oil derivatives. The driving factors of these massive initiatives are the adverse consequences of petroleum-based products, petroleum depletion, rising crude oil prices, and, most importantly, the additional benefits of renewable biomasses. The long flexible fatty acid structure behaves as the perfect soft segment for polyurethanes (PUs), and the unsaturation in the chains acts as reaction centers for new structural design. Vegetable oils are a favorite of industrialists and researchers due to their renewable nature, ease of accessibility, capacity to make various structural alterations, low cost, and absence of side effects. The leading plant oil derivatives include epoxy resins, polyurethanes, and polyesters. Because of epoxies' growing health and environmental concerns, the public has embraced vegetable oil-based waterborne polyurethanes the most [19].

The latest advances in bio-based self-repairing films are addressed, with a focus on corrosion resistance. Yixi Chen et al. created a self-healing polymer coating by adding 10% microencapsulated tung oil, as a healing agent, to conventional epoxy resin. The microcapsules broke when the coating was damaged, releasing tung oil that cured throughout the damaged area. When compared with mild steel samples with conventional and self-healing coatings in accelerated corrosion tests, samples with self-healing performance were at least three times better. Eighty-three percent of the samples showed no corrosion after 150 days of accelerated corrosion examination. Pull-out testing exhibited similar bond stresses as that of normal epoxy coatings [20].

In another significant study, multifunctional waterborne polyurethanes from castor oil with high strength were successfully synthesized by Chaoqun Zhang et al. using dithiodiphenylamine (DTDA). The polymer matrix exhibited high tensile strength (up to 38 MPa) and self-repairing properties. The films' diverse shape-memory effects, including dual-to quadruple shape memory, were enhanced by the samples' wide glass transition temperature. Castor oil-based WPU has self-healing properties because of the numerous H-bonds and urethane bonds [19].

Microcapsules, with biopolymers derived from cashew nuts and gum arabic, were filled via interfacial polymerization. The Q235 steel substrate was protected with an epoxy polymer embedded with these microcapsules. The effectiveness of this composite coating was determined by immersing coated scratched and coated samples without a scratch in seawater, and analyzing them with a scanning electron microscope (SEM). The corrosion inhibition efficiency and self-healing property of the polymer composite were assessed using surface analytical (SEM) and electrochemical impedance spectroscopy (EIS) techniques. Both polymers suppressed corrosion in the absence of a catalyst and showed excellent self-healing properties. With the combined effect of both polymers matrices in a composite, the effect was more pronounced [21].

Zheludkevich et al. developed a superior self-healing coating by including a potent inhibitor, such as cerium ions, in a pre-coating of chitosan as a reservoir. The combined result exhibited good corrosion inhibition and self-healing in a localized electrochemical assay [22]. Neem oil, linseed oil, and cardanol are some bio-based materials with self-healing properties that are currently being researched. Another study discovered that neem oil has exceptional self-healing properties. Another potential mechanism is oxidation in atmospheric oxygen. However, it is barely understood and is thought to be a radical initiated mechanism [23]. The efficiency of linseed oil-filled microcapsules for the repairing of paint/coating scratches was examined. When linseed oil was discharged under simulated mechanical action, cracks in a paint coating were successfully repaired. Linseed oil healed the region and prevented the substrate from corroding [24]. Samadzadeh et al. investigated the efficacy of encapsulated tung oil. EIS and immersion tests were used to assess corrosion resistance in the healed portion. Tung oil released from shattered microcapsules successfully repaired the artificial crack in the coating [25]. FE-SE images of a ruptured microcapsule containing linseed oil are shown in Figure 8.5.

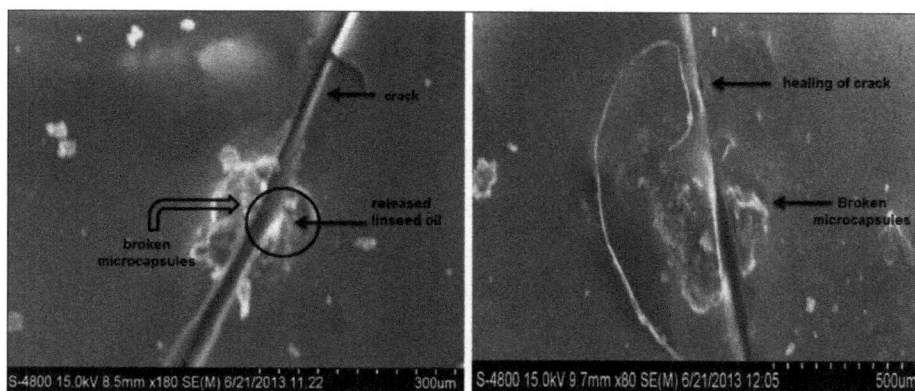

Figure 8.5: FE-SEM images of the Self-healing effect of linseed oil from ruptured microcapsules (adapted with permission from reference [23]. Copyright A.B. Chaudhari, P.D. Tatiya, R.K. Hedaoo, R.D. Kulkarni, V. V Gite, some rights reserved; exclusive licensee American Chemical Society. Distributed under a Creative Commons Attribution License 4.0 (CC BY) https://creativecommons. org/licenses/by/4.0/).

Khorasani et al. studied the bio-based self-healing technologies that have been published so far and discovered that coatings with vegetable oils could be made by in-situ polymerization in an oil-in-water emulsion. This process is costly, time-consuming, produces a large number of by-products, and has low encapsulation yield and reaction efficiency, making it unsuitable for large-scale applications. Electrospray encapsulation can fix the major synthesis problem, and incorporating a glycidyl group can speed up the healing process [26].

8.3.3 Self-cleaning coatings

Using either hydrophilic or hydrophobic strategies, nature-inspired self-cleaning coatings based on surface contact angles provide an autonomous action of dirt removal from a surface. Two different approaches have been used to investigate self-cleaning surfaces. The first approach includes putting a photocatalytic coating on the substrate surface, which uses ultraviolet light from the sun to catalytically break down organic dirt. At the same time, the surface becomes superhydrophilic, distributing water evenly across the coating and causing dripping, to leave fewer drying traces. The lotus effect phenomenon, where the surface becomes superhydrophobic, is the idea behind the second self-cleaning method. The contact angle of water determines the degree of hydrophobicity. Hydrophobic surfaces have contact angles greater than 90°, while superhydrophobic surfaces have contact angles up to 150° or more and repel water droplets completely [27]. The SEM images of different self-cleaning surfaces and their self-cleaning mechanisms [28] are shown in Figure 8.6.

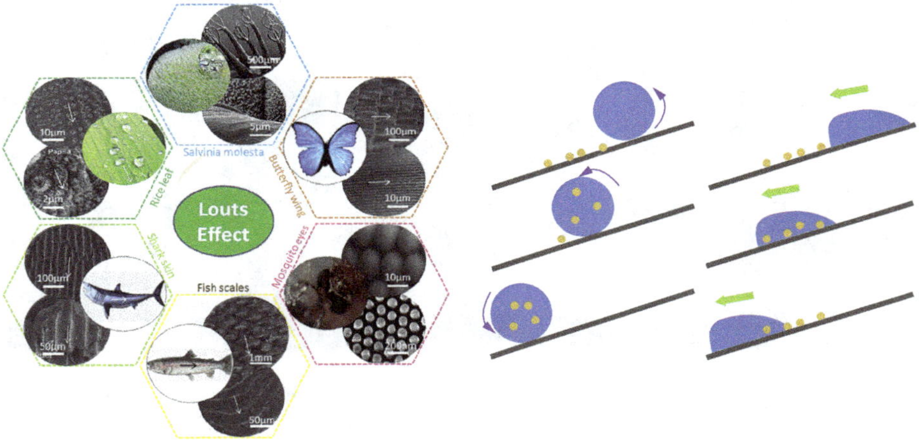

Figure 8.6: (i) Self-cleaning surfaces present in nature and corresponding Scanning Electron Microscope images (ii) Lotus effect (a) self-cleaning and (b) wetting based on contact angle differences (adapted with permission from reference [28]. Copyright M. Zhang, S. Feng, L. Wang, Y. Zheng, some rights reserved; exclusive licensee Elsevier. Distributed under a Creative Commons Attribution License 4.0 (CC BY) https://creativecommons.org/licenses/by/4.0/).

By roughening the surface of a hydrophobic surface, superhydrophobic coatings can be created. The self-cleaning action of a coating depends on the sliding angle (SA) and the water contact angle (WCA); the water drop should slide off from the surface without leaving any trace [29]. Superhydrophobic surfaces (WCA > 150° and SA < 10°) have potential self-cleaning, antifogging, anticorrosion, and oil/water separation applications. Figures 8.6 show some of the naturally occurring self-cleaning surfaces and their SEM images.

Self-cleaning anticorrosive materials, with superhydrophobic coatings, are incredibly inspiring candidates for improving corrosion performance. In another investigation led by Cheng et al., superhydrophobic polyurethane coatings were prepared from waste cooking oil. The polyurethane prepolymer was developed from waste cooking oil and modified with amino-terminated polydimethylsiloxane (ATP) to obtain a super-hydrophobic emulsion. Among the various composites of ATP and silicon carbide, the coating with 20 wt% of particles in polyurethane showed good pull-off strength and thermal conductivity. In addition to the self-cleaning property, the waste oil-based polyurethane coating has good wear resistance, and its corrosion inhibition performance can meet the requirements of air coolers in special environments. Figure 8.7. shows the self-cleaning action of superhydrophobic polyurethane [30]. The WCA and SA are 161° and 3°, respectively.

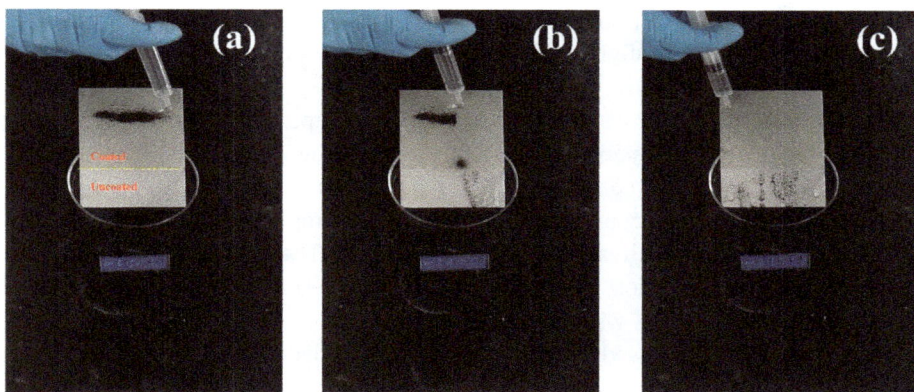

Figure 8.7: Self-cleaning process of the superhydrophobic PU coating. (**a**) Before the analysis, (**b**) During the analysis, (**c**) after the analysis (adapted from reference [30]. Copyright Y. Cheng, D. Miao, L. Kong, J. Jiang, Z. Guo, some rights reserved; exclusive licensee MDPI (Multidisciplinary Digital Publishing Institute) Distributed under a Creative Commons Attribution License 3.0 (CC BY) unported licence https://creativecommons.org/licenses/by/3.0/).

Atta et al. prepared a self-cleaning superhydrophobic coating, with self-healing and anticorrosion properties, from a cardanol derivative. Firstly, a Schiff-base polymer was made. Cardanol was extracted from cashew nut oil and modified using a bishydrazino-s-triazine derivative, followed by condensation polymerization or reaction with terephthaldehyde. The multifunctional cardanol monomer and polymer were introduced to an epoxy/polyamine hardener solution in various weight percentages. After more than 1,500 h of exposure, all coatings demonstrated better thermal, mechanical, adhesion, and anticorrosion properties than standard epoxy coatings. The 3 wt. % coating has the best self-cleaning and self-healing properties. Coatings increased the contact angle of seawater by more than 150° [31].

The modification of cardanol glycidyl ether with pentaethylenehexamine resulted in hydrophobic polyamine hardeners (PEHA). To make hydrophobic and superhydrophobic epoxy coatings, bisphenol A diglycidyl ether (DGEBA), cured by dicardanoxy (DCHI) and capped nanoparticles were coated on a rough steel surface. Magnetite nanoparticles (Fe_3O_4 NPs) were capped with this to produce superhydrophobic magnetic nanoparticles. The coating contained DCHI-Fe_3O_4 nanoparticles with a WCA > 150°, according to the seawater contact angle measurements. Excellent adhesion, mechanical, and anticorrosion performances were achieved in salt spray fog analysis of 2,000 h [32].

A superhydrophobic and self-cleaning photo luminescent protein coating was prepared by Xie et al. by incorporating hydrophobic silica nanoparticles in soy protein with some simple chemical reaction steps. The film exhibited fluorescence emission on UV irradiation and also resisted wetting by most fluids, showing superhydrophobic and self-cleaning properties. It could resist acid/alkaline corrosion, solvent immersion, and slight mechanical abrasion without losing its basic functional properties [33].

Zheng et al. prepared water-based isosorbide bio-epoxy resin with outstanding hydrophobicity and corrosion protection to eliminate the water absorption and brittleness of standard epoxy coatings without the use of organic solvents. Spin coating was used to coat this epoxy and incorporated superhydrophobic SiO_2 NPs, with a hydrophobic hardener and hexadecyltrimethoxysilane. The coating demonstrated a positive change in corrosion potential, from −1.008 to −0.747 V, and a considerable decrease in the i_{corr} value, with 92% inhibition efficiency in electrochemical experiments [34]. Zhong et al. used 2-mercaptoethanol-modified castor oil to prepare self-cleaning coating with anti-graffiti and antifingerprint properties [35]. Liang et al. also used epoxidized castor oil to synthesize a corrosion-resistant self-cleaning coating with an organic-inorganic hybrid system with UV curing [36].

8.3.4 Antifouling coatings

The undesirable accumulation and proliferation of plants, microorganisms, animals, and dirt on surfaces immersed in water are known as biofouling. Each year, biofouling costs billions of dollars to the marine, shipping, and other global industries, and cannot be ignored. Micro-fouling and macro-fouling are two types of marine fouling. Corrosion of metallic parts is a detrimental effect of biofouling, destroying structural components and disrupting the system's overall functioning. Specially formulated surface coatings are utilized to prevent the surfaces from biofouling. Two types of antifouling coatings are used to reduce biofouling and its impacts. The first is biocide-containing coatings, while the second is biocide-free coatings. Chemically active self-polishing coatings

(SPC) contain booster biocides, and fouling release coatings (FRC) with silicone and fluorine coatings eliminate biofouling by reducing the surface energy and elastic modulus. This allows the foulant to be easily removed through simple mechanical cleaning or water movement. However, FRC flops to inhibit biofilm proliferation but inhibits attachment of most macro fouling in dynamic conditions [5, 37]. Major anti-fouling coating approaches in the marine environment are given in Figure 8.8.

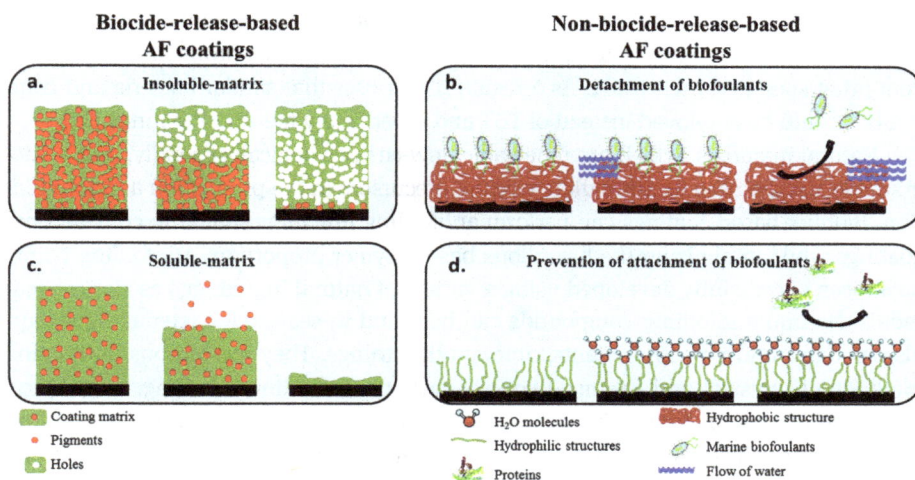

Figure 8.8: (a–d). Antifouling coating approaches in the marine environment (adapted from reference [37]. Copyright Nurioglu AG, Esteves AC, Gijsbertus de With, some rights reserved; exclusive licensee Royal society of Chemistry. Distributed under a Creative Commons Attribution License 3.0 (CC BY) unported licence https://creativecommons.org/licenses/by/3.0/).

Marine biofouling is of great concern because of the substantial economic losses associated with micro and macro colonizing on the shipping hulls, which reduce fuel efficiency and dry-docking operations. Fouling formation has many detrimental effects, but the most important ones are as follows;

- Increase in fuel consumption by 30–40%
- Air pollution by emission of greenhouse gases
- Transportation of invasive species
- Destruction of the hull surface
- Hull maintenance
- Damages to underwater structures
- Biocorrosion

These factors result in a massive rise in pollution as well as financial burden. Anti-fouling coatings should have high antifouling properties as well as good adherence and durability. They should be economical and readily available with corrosion inhibition property, smoothness, and fast-drying on application [38, 39].

Tributyltin and copper compounds have been employed as effective antifouling agents in coatings for decades; however, copper compounds, TBT, and their derivatives have an adverse environmental impact and are not biodegradable. To avoid these issues, extensive research and development have been carried out in the field of renewable, eco-friendly, and bio-based antifouling paints and compounds. Many seaweeds and marine vertebrates are found to be devoid of biofilms. This is due to the release of chemicals created by the bacteria that live on them into the environment, which prevent other species from settling. As a result, it proposes that similar bacteria and compounds could be employed instead of TBT and other heavy metal compounds [39].

Natural materials, which are renewable, low-cost, ecologically friendly, and biodegradable, can be used as antifouling coating precursors in the production and formulation. Biofiber-based coatings can perform antifouling functions similar to conventional coatings while minimizing the hazardous base polymer proportion. Antifouling paints have been successfully developed using a variety of natural ingredients as starting materials. Natural antifouling compounds can be found in seaweed, bacteria, algae, corals, sponges, and terrestrial plants, among other things. They use various antifouling methods of physical and chemical control systems, including low drag, wettability, low adhesion, microtexture, chemical secretion, grooming, and sloughing. Tannins are large polyphenolic compounds, which have antimicrobial and anticorrosion potential. *Andrographis paniculata* extract and isolated compounds exhibited excellent antibacterial action and antibiofouling in marine conditions [38].

Jiansen et al. used a thiol-ene reaction and polyaddition to create a bio-based poly(lactic acid)-based PU with hydrolyzable triisopropylsilyl acrylate (TSA) functionality. The coating had high adhesive strength (2.0 MPa) as well as a slow degradation rate. The polymer coating contains an environmentally benign antifoulant butenolide, obtained from marine bacteria, which offers a controlled release when the polymer degrades. Digital holographic microscopy tracking analysis showed the effective antifouling action of the coatings in seawater against Pseudomonas sp. and other marine foulants for more than three months. Within one month, the entire surface of the control was covered with diverse species, primarily tubeworms, bryozoans, and algae, showing significant fouling in the area. On the other hand, the surface coverage of the bio-based surface was low even after three months [40].

In another significant study, Somisetti et al. synthesized a natural product-based antifouling polyurethane coating from undecylenic acid, a by-product of castor oil and phosphated cardanol-based polyol. The anticorrosion and antibacterial properties of the coating were improved by combining the effects of phosphorus and bio-based polyol with sulfur and nitrogen. For the prepared polyurethane coatings, the corrosion rate was reduced dramatically, from 3.010^{-4} to 1.810^{-7} mm per year (Figure 8.9) [41]. The Tafel plots of hydroxylated cardanol (CDPOH) and phosphate cardanol (CDPOH)-based polyurethanes are given in Figure 8.9.

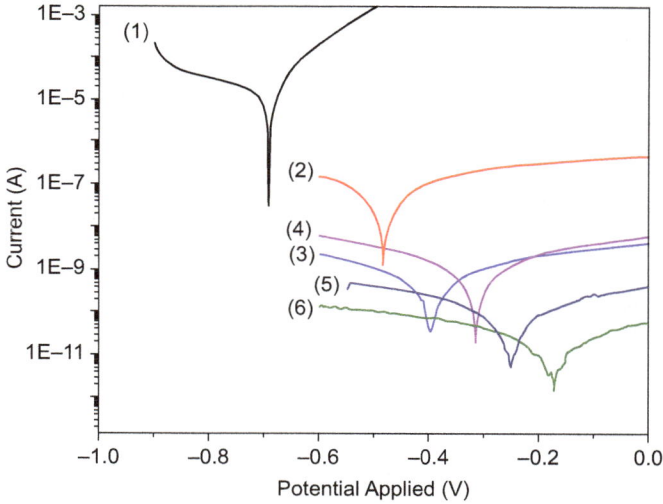

Figure 8.9: Tafel plot of (1) Bare MS panel, (2) CDOH-PU, (3) CDPOH-PU, (4) CDPOH-5% PU, (5) CDPOH-15% PU, and (6) CDPOH-25% PU in 3.5% NaCl (adapted with permission from reference [41]. Copyright V. Somisetti, R. Narayan, R.V.S.N. Kothapalli, some rights reserved; exclusive licensee Elsevier. Distributed under a Creative Commons Attribution License 4.0 (CC BY) https://creativecom mons.org/licenses/by/4.0/).

Natural lacquers have long been utilized as renewable and environmentally beneficial coatings. Chen et al. created a novel fast curing hybrid lacquer with high adhesion, high barrier performance, corrosion inhibition, excellent mechanical properties, flexibility, and good hardness. Because of the self-enrichment of the amphiphilic telomer, the coating had outstanding fouling-resistant performance against diverse microorganisms and is shown in Figure 8.10 with the help of a fluorescence microscopy images [42].

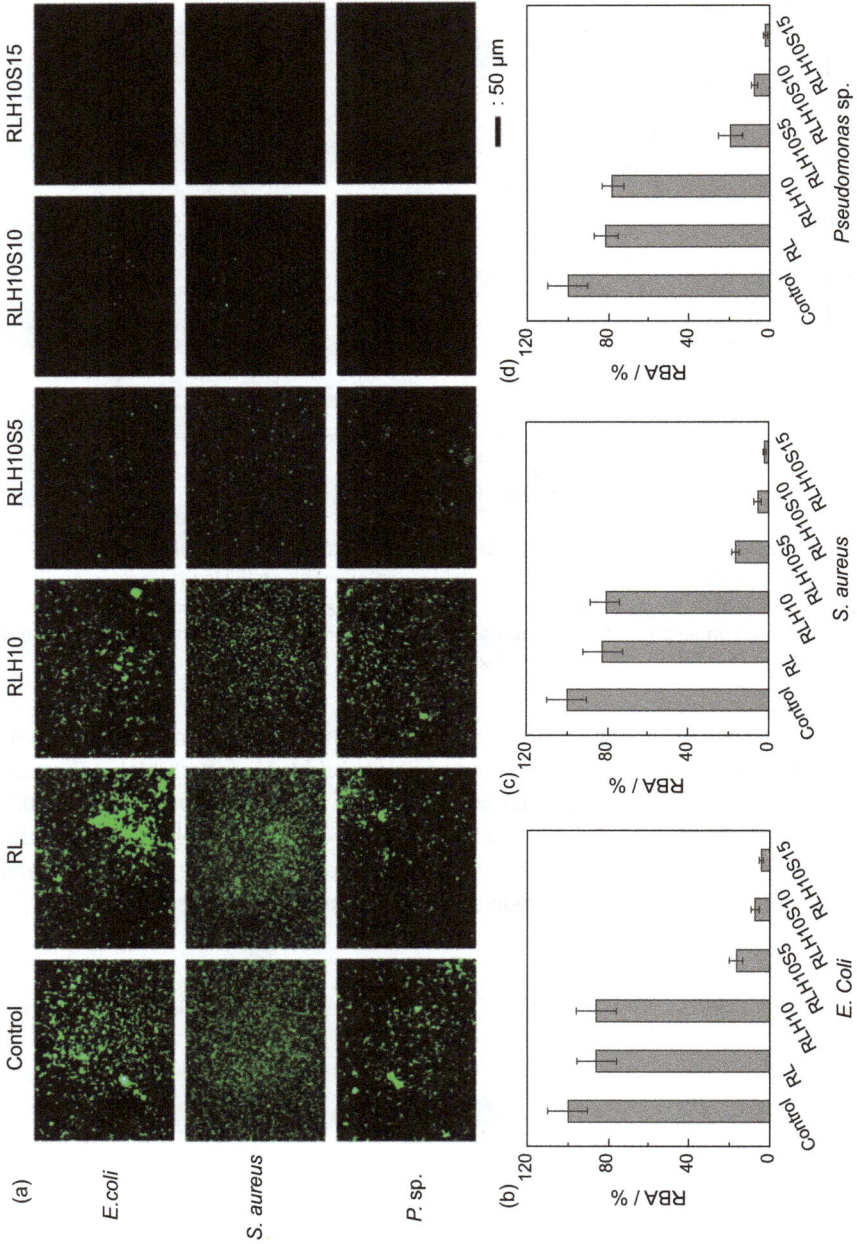

Figure 8.10: (a) Fluorescence microscopy images of microorganisms attached to the hybrid lacquer layer when cultured for four hours. Relative bacterial adhesion of (b) *E. coli*, (c) *S. aureus*, and (d) marine bacterial *Pseudomonas* sp. (adapted with permission from reference [42]. Copyright Y. Chen, G. Zhang, G. Zhang, C. Ma, some rights reserved; exclusive licensee Elsevier. Distributed under a Creative Commons Attribution License 4.0 (CC BY) https://creativecommons.org/licenses/by/4.0/).

8.3.5 Antimicrobial coatings

According to a recent study published in the New England Journal of Medicine on March 17, 2020, the coronavirus causing the current pandemic can survive for up to three days on stainless steel and plastic surfaces. According to the study, the virus can survive in the air for up to 3 h, on cardboard for up to 24 h, on copper for up to 4 h, and on plastic and stainless steel for up to 72 h. Many viruses and bacteria can survive for extended periods in the air or on surfaces. They can transmit disease through public transportation, hospitals, and other high-touch surfaces in public places, such as floors, walls, and doorknobs. As a result, it is critical to focus on developing coatings that may kill microorganisms by contact mechanism. The method of action of antimicrobial surface coatings is shown in Figure 8.11.

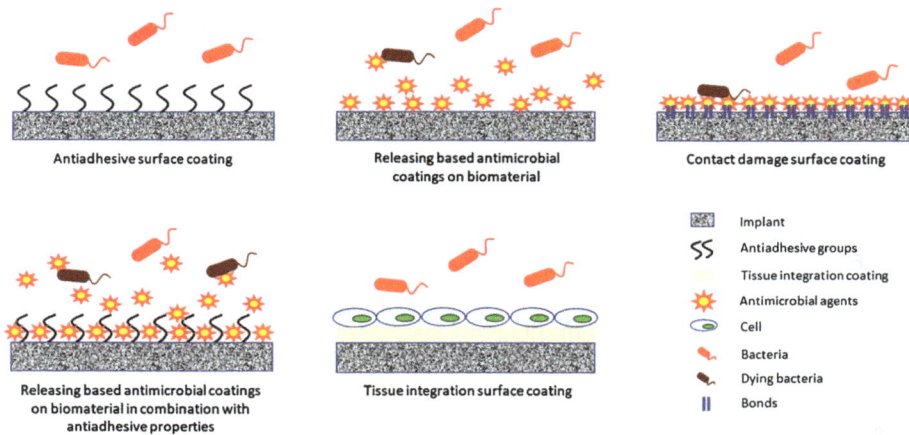

Figure 8.11: Different methods of action of antimicrobial surface (adapted with permission from reference [43]. Copyright X. Li, T. Huang, D.E. Heath, N.M. O'Brien-Simpson, A.J. O'Connor, some rights reserved; exclusive licensee AVS. Distributed under a Creative Commons Attribution License 4.0 (CC BY) https://creativecommons.org/licenses/by/4.0/).

Deacetylation of chitin generates chitosan from crustacean shell waste. Chitosan is used to make a variety of nanocomposites and protective coatings. It has antibacterial and antifungal properties. The interaction between the positively charged amino groups from the chitosan coating and the negative charge on the bacteria's cell membrane, which finally leads to the leaking of cellular components, is the primary explanation for the antibacterial effects. As a chelating agent, chitosan selectively interacts with metals to inhibit microbial development (enzyme deactivation). Chitosan forms an impenetrable polymeric layer that prevents nutrients from reaching the cell, inhibits protein synthesis, etc. – the primary mechanism of action of antimicrobial behavior. Chitosan-based coatings are also reported to have excellent corrosion inhibition and self-healing properties [44].

In another significant study, Zhang et al. synthesized vegetable oil-based cationic waterborne polyurethane with long-term antimicrobial coatings, showing anticorrosion performance from castor oil and aminoacids [45]. Ulaeto et al. loaded antimicrobial phytochemical Quercetin in mesoporous silica nanoparticles and dispersed in epoxy. The coatings on Al-6061 alloy showed high corrosion inhibition efficiency and antimicrobial action against *Pseudomonas nitroreducens* [46] (Figure 8.12).

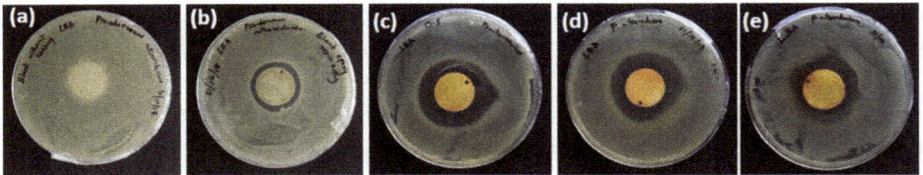

Figure 8.12: Antimicrobial activity of the epoxy layer against Pseudomonas nitroreducens; (a) Control – bare Al (b) unloaded epoxy on Al, (c–e) 0.5, 1.0, 3.0 wt% quercetin composite-epoxy coatings showing the zones of inhibition (adapted with permission from reference [46]. Copyright S.B. Ulaeto, A. V Nair, J.K. Pancrecious, A.S. Karun, G.M. Mathew, T.P.D. Rajan, B.C. Pai, some rights reserved; exclusive licensee Elsevier. Distributed under a Creative Commons Attribution License 4.0 (CC BY) https://creativecommons.org/licenses/by/4.0/).

8.3.6 Fire-retardant coatings

Intumescent coatings and non-intumescent coatings are two classes of fire-retardant coatings. Intumescent paints or coatings are used to passively fireproof a structure by slowing the passage of heat from the fire to the internal structure through the surface. This ensures the structural integrity of the buildings and gives us more time to evacuate. Intumescents are utilized in applications, such as fire stopping, fireproofing, gasketing, and window casings to provide passive fire protection. Heat exposure causes an intumescent coating to swell, resulting in increased volume and decreased density. They can be used as fireproofing paint on structural parts. Many chemicals are used in intumescent paint, including an acid source, a charring agent, a blowing agent, and a binder. A vital component of an intumescent coating is the binder. Its primary function is to bind all molecules together, but it also serves as a carbon source and impacts the foaming process. The role of various binders, as well as their chemical reactivity, has been studied. When a blend of a linear copolymer with solid reactivity, with the acid source and a crosslinked copolymer as the binder in the intumescent paint was used, the thermal insulation greatly increased [47]. From the formulations of water-soluble salts containing phosphate, borax, sulfamate, nitrate, sulfate, boric acid, and halides, the active non-intumescent films discharge fire retarding agents, including gas-phase radicals [48].

Residential areas, business plots, offshore structures, vehicles, ships, and aircraft, all benefit from these uses. These coatings protect the wood, textile, and polymer surfaces more than metal surfaces. In case of a combination of such matrices, the corrosion-resistant coating can operate as an intelligent coating when exposed to an external heat stimulus. The choice of suitable flame-retardant fillers affects the characteristics of the formulation. It is usually better to give a material with good corrosion resistance and mechanical properties, a fire-retardant property, than vice versa.

Intumescent coatings with different percentages of coconut fiber (CCN), wood waste (MDP), and peach stones (PEA) biomasses were investigated. CCN and MDP had an ideal dry mass percentage of 9%, while PEA had an ideal dry mass percentage of 6%. After fire resistance testing, the coatings expanded by additional 600% in the case of CCN9, 1,300% in the case of MDP9, 1,500% in the case of blank, and 1,600% in the case of PEA6. The highest temperatures on the back of the material for biomass-based intumescent coatings were 120 °C, compared to 474 °C for uncoated steel [8]. Cardanol is classified as inedible oil, making it an excellent renewable source of feedstock for producing chemicals and materials for various purposes. Although brominated cardanol derivatives have strong fire resistance, brominated derivatives harm the environment. Adding phosphorus atoms to bio-based polymers improves their fire resistance, metal adhesion, and anticorrosion properties significantly [49].

Polypropylene (PP) is a widely used material with superior mechanical strength, electrical insulating property, and corrosion inhibition, but it is highly flammable. To overcome this defect for the application in some fields, Wang et al. synthesized three bio-based salts from three different amino acids, viz. lysine, arginine, and histamine, with phytic acid. The flame-retardant efficiency was analyzed using various methods; in Underwriters Laboratories test standard UL 94 (UL-94) and Limiting oxygen index study (LOI), it was found that PaArg > PaLys > PaHis; but in cone calorimeter test, the performance was as follows: PaLys > PaArg > PaHis. The efficiency depends on the char forming rate and strength of the formed char. In this aspect, PaArg and PaLys are superior to PaHis. The second factor affecting the efficiency is the diluting action of the released water, carbon dioxide, and the quenching effect of phosphorous-containing radicals. In this aspect, PaLys has a specific advantage under combustion with a continuous radiation source. This study hopes that using entirely bio-based phytic acid-basic amino acid salt as flame retardant will enable the use of polypropylene with a fire-safety function and outstanding mechanical qualities in much likely fire-causing scenarios [50]. To improve the flame-retardant efficiency of PP, Wang et al. employed different phytic acid/piperazine salts (PHYPI). PHYPI displayed good intumescent behavior with a molar ratio of 1/3 (phytic acid/piperazine) [51].

Zhang et al. created a bio-based epoxy formulation with exceptional flame-retardant properties from itaconic anhydride. Low phosphorus levels considerably increased both the toughness and flame retardancy of epoxy resins in gas and condensed phase flame-retardant mechanisms [52]. A novel ricinoleic acid (RA)-based phosphorus and nitrogen-bearing flame-retardant polyols were successfully formulated by Chu et al. [53]. Polyurethane coatings are well-known for their corrosion resistance, but they are also flammable. Fan et al. used melamine starchphytate (PSTM) to turn polyurethane coatings into flame-retardant coatings. PSTM promotes thorough charring of PU, which prevents the foam from combustion [54].

8.3.7 Anti-graffiti coatings

Graffiti poses a severe threat to heritage materials and damages a wide range of surfaces, resulting in expensive cleaning. In many cases, graffiti penetrate the holes in the substrate, causing permanent damage to the painted area. Anti-graffiti coatings can be prepared by incorporating particles repellent to both water and oil. The resulting surface will be an easy-to-clean surface that can survive repeated graffiti attacks. Fluorinating agents are an excellent choice for developing anti-graffiti coatings, which will help lower the surface energy by migrating to the surface and preventing the marker paints from adhering [5]. However, the development of a bio-based product is necessary for the prevention of pollution by fluorinated products. Most of the anti-graffiti products available in the market today are siloxane/silicone-based, which are highly hydrophobic and repel water-based pen and markers [55].

Naturacoat is a sacrificial coating technology developed by Beardow Adam's that offers complete protection against graffiti and pollution, making it ideal for historical and high-profile monuments and locations. It is a paint formulation made from natural polysaccharides and water. It is a sacrificial coating that may be applied by a simple spraying approach. This coating can be easily cleaned if it becomes contaminated or if graffiti is spotted, by high-pressure warm water around 70 °C. This can be applied to natural stone, brickwork, metals, plastics, ceramics, and marbles [56].

8.3.8 Antismudge coatings

Antismudging coatings are used on various substrates, including plastics, glass, metals, and textiles. They have scratch and abrasion resistance and antistatic, easy-to-clean, antireflective, and antimicrobial qualities. Antismudge technology solutions reduce the likelihood of smudges forming and make cleaning easy, improving customer satisfaction. A coating that repels dirt, oil, and water, reduces fingerprints and makes it easier to clean electronics display surfaces. It is perfect for phones, tablets, watches, displays, and monitors, among other things. Stains on a building's glass curtain wall also impact its aesthetics and lighting, demanding manpower to clean in high altitude, resulting in high costs and safety risks.

Surface morphology, surface chemistry, and interaction with lubricating components are just a few of the physicochemical properties that can be tweaked to improve antismudge capabilities. Gas cavity generation is promoted by surfaces with modified microstructures and increased roughness. This implies that the solid/liquid interface is partially replaced by a gas/liquid interface (Cassie-Baxter state) for liquids in contact with such a surface, reducing the liquid/solid contact area and giving a super-omniphobic surface, with WCA greater than 150° and a very low SA.

Reacting acetylated starch as the polyol and hexamethylene diisocyanate trimer (HDIT) as the hardener, Lei et al. created a bio-based antismudge polyurethane coating. Coatings with excellent transparency (>97%) and outstanding antismudge qualities can be made by changing the polydimethylsiloxane (PDMS) and starch content. Liquid pollutants, such as inks, water- and oil-based substances, can be easily cleaned from the coated surface by "sliding-off," "beading," and "wiping-off" without leaving any traces and marks. The coating's corrosion inhibition performance on a metal substrate was also tested, and it was discovered to be an excellent corrosion inhibitor. As a result, bio-based "green" coatings with omniphobic qualities have the potential to protect surfaces against contamination and corrosion, which is crucial for the long-term development of functional coating materials [57].

8.3.9 Anti-icing coating

Salts, both solid and liquid, are employed in winter for road and surface maintenance because they are good at breaking and preventing the bonding of snow and ice to road surfaces and roofs. Due to the negative impact of salt on buildings and the environment, many alternative materials are being researched for snow and ice control during winter.

Hydrolysis of starch and other polysaccharides into glucose, followed by hydrogenation and other chemical transformations, can be used to make isosorbide. Zheng et al. created an environmentally benign waterborne hydrophobic bio-epoxy coating by combining dual-scale SiO_2 nanoparticles with (3-glycidyloxypropyl) trimethoxysilane in an aqueous combination containing an isosorbide-based epoxy and a hydrophobic hardener. On an iron foil, a spin-coated sample had a WCA of $153.0 \pm 1.1°$ and a SA of $14.3 \pm 1.9°$. After sandblasting, it demonstrated outstanding self-cleaning properties and very low dirt accumulation. It exhibited much lower icing temperature, a significantly longer icing delay time, and low ice adhesion strength at 0.101 ± 0.019 MPa. The new coating is bio-based and environmentally friendly, with potential uses in marine, aircraft, energy harvesting, and sports [58].

Zhong Chen et al. prepared cardanol-derived resins with furfurylamine (FA) or 1,8-diamino-*p*-menthane (DAPM) as the hardener in solvent-free conditions. The resulting surface exhibited very low ice adhesion strength at 55.0 ± 5.2 kPa. It also has good anticorrosive capabilities, as seen by its low i_{corr} and high E_{corr} values. The ice adhesion strength was significantly reduced after PDMS hydrophobization [59].

Ice-phobic surfaces are more practical than anti-icing surfaces in real life. Traditional epoxy coatings were changed by Feng et al. by adding maleic anhydride as a curing agent, as well as a small amount of epoxy resin grafted with fluorine-containing chains. The wettability, adhesion strength of ice on the coated surface, and the stability of the coating's ice-phobic qualities were investigated in depth. The modified coating surface became more hydrophobic, requiring less force to remove ice than the unmodified coating surface [60]. The coating surface keeps its hydrophobic and ice-phobic qualities after sanding with abrasive paper. The absence of polar groups is most likely to be responsible for the coating's exceptional and sustainable ice-phobic activity.

8.4 Corrosion inhibition mechanism of natural product-based smart coatings

There are different mechanisms behind the corrosion inhibition action of natural product-based inhibitors. Some of them may act as cathodic inhibitor, anodic inhibitor, or mixed-type inhibitor. Lowering the rate of diffusion for reactants to the metal surface or by lowering the electrical resistance of the surface is another technique [61]. Some other compounds work by their property to make the surface super-hydrophobic to minimize the interaction between the surface and the water molecule. Some coatings have a self-healing ability that cures the scratches on the surface and prevents further corrosion into the metal lattice. The self-cleaning coatings aid in the anticorrosion mechanism by removing the dirt and other impurities from the surfaces, which may otherwise trigger corrosion. The presence of antimicrobial compounds in the coatings can inhibit the corrosion induced by the metabolites of microbes.

The corrosion inhibition potential of the molecules can be attributed to the coordination of the lone pair of electron to metal. It may be due to a single compound in the extract or it is also possible to be the synergic effect of many molecules present in the system [61]. The action of many inhibitors is by forming a physisorbed or chemisorbed protective film on the metal surface via polar functional groups containing –N, –S, –O, and/or conjugated multiple bonds and inhibit the corrosion-prone surface site [62]. There are a large number of heterocyclic phytochemicals identified in the plant and animal kingdom.

Xin Lai et al. depicted the mechanism of corrosion resistance of green inhibitors by taking chitosan derivative as an example (Figure 8.13). The major interactions between the inhibitor and the metal involve chemical and physical adsorption, hydrogen bonding, and retro donation [63].

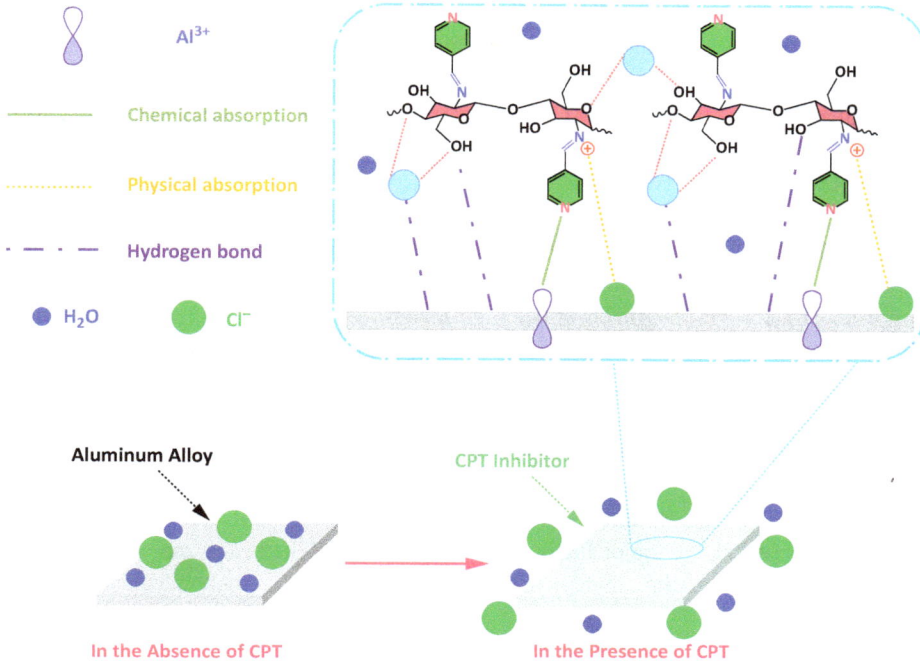

Figure 8.13: The mechanism of action of chitosan derivative (CPT) in corrosion inhibition (adapted with permission from reference [63]. Copyright X. Lai, J. Hu, T. Ruan, J. Zhou, J. Qu, some rights reserved; exclusive licensee Elsevier. Distributed under a Creative Commons Attribution License 4.0 (CC BY) https://creativecommons.org/licenses/by/4.0/).

Dahmani et al. studied the anticorrosion property of cinnamon essential oil for copper in 3.0 wt% of saline medium and identified the constituent responsible and the mechanism of action with the help of density functional theory calculations and molecular dynamic studies. The GC-MS analysis reveals that cinnamon oil contains many molecules, such as α-phellandrene, D-3-carene, α-terpinene, *p*-cymene, limonene, 1,8-cineole, phenylacetaldehyde, γ-terpinene, acetophenone, terpinolene, *trans*-β-ocimene, and α-thujone, β-thujone. The amount of electronic transition (ΔN) from the substituent to the metal atom is a key indicator of anticorrosion activity. In recent corrosion inhibition investigations, two novel parameters developed by Gazquez and colleagues, namely electron-donating power and electron-accepting power, have received a lot of attention. These values can be correlated to the HOMO-LUMO concept of Koopmans Theorem, which gives an idea about electron affinity and ionization energy. The MD method can easily anticipate the strength of the interaction between the inhibitor compounds and the metal surfaces. DFT calculations and MD simulations for *trans*-cinnamaldehyde, δ-cadinene, and β-cubebene molecules were carried out among these identified compounds. All three molecules have the planar structure in the optimized geometry, which ensures maximum interaction with the surface. The energy of HOMO also makes some comments regarding the ability of molecules to donate electrons to a low-energy accepting species. The existence of –O atom and π-electrons determines the extent of electron donation in cinnamaldehyde and its interaction with metals. The presence of π-electrons determines the inhibition efficiency of the other two molecules [62].

The important quantum chemical parameters derived from DFT calculations, such as EHOMO, ELUMO, energy gap ΔE, polarizability α, electrophilicity ω, nucleophilicity ε, ΔN, and electronegativity are the factors controlling the anticorrosion property of all the molecules and the higher rate for cadinene is its correct order of these parameters. Cadinene has lower electronegativity, which implies that it tends to discharge its loosely held electrons more easily than the other two compounds. The trend of polarizability indicates that cadinene is easy to polarize than the other two. The order of electrophilicity ω values indicates that cadinene has a less tendency to accept the electrons and it is effective against corrosion. The MD simulation showed that the energy of adsorption shows a lower value for cadinene on Cu (Figure 8.14). Both studies reveal the major factors and depending on them, the major inhibitor is cadinene, followed by cinnamaldehyde along with all other phytochemicals [62].

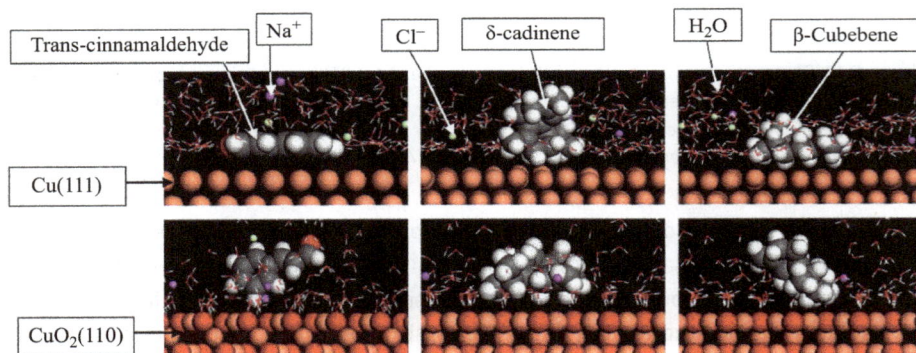

Figure 8.14: Configurations of *trans*-cinnamaldehyde, δ-cadinene, and β-cubebene molecules adsorbed on Cu(111) and CuO$_2$(110) in acidified 3.0 wt% saline medium (adapted with permission from reference [62]. Copyright K. Dahmani, M. Galai, M. Ouakki, M. Cherkaoui, R. Touir, S. Erkan, S. Kaya, B. El Ibrahimi, some rights reserved; exclusive licensee Elsevier. Distributed under a Creative Commons Attribution License 4.0 (CC BY) https://creativecommons.org/licenses/by/4.0/).

In another important study, Emeka E. Oguzie et al. shows that a strong chemisorption force was responsible for the interaction between the inhibitor and the metal surface. They used the chemisorptive interactions between the piperine molecule, the active and primary component of the piper guineense extract, and the Fe surface, and theoretically modeled using DFT-based quantum chemical computation.

The magnitude of the measured adsorption energy shows the piperine molecule's significant chemisorption, confirming the experimental results [64].

8.5 Applications and commercial viability of bio-based smart coatings

Medical and healthcare, oil and gas, aerospace, packaging, automotive, consumer electronics, construction, marine, military, energy, textiles, and apparel industries, etc. are among the end-user applications of intelligent coatings. Stimuli-responsive coatings can be designed to suit the needs of individual customers [5]. Easily contaminated surfaces, such as motor cars, building walls, fabrics, paper, metals, glass, solar panels, and wooden surfaces, can benefit from superhydrophobic self-cleaning coatings to protect them from dirt collection and corrosion attack [65]. These self-cleaning and antismudge coatings save the use of cleansers and the time paid to clean everything, from personal computers to kitchen surfaces to shower screens.

Antifingerprint coatings are in high demand in the automobile industry, thanks to the increasing use of touch panel screens in vehicles. In automotive displays, scratch-proof coatings, antifingerprint mirrors, interior surface coatings, wear and corrosion-resistant coatings for engines, lubricant additives, self-cleaning coatings on glass, anticorrosion engine coatings, antimicrobial interior trim, and upholstery coatings, hydrophobic and oleophobic, antifingerprint coatings are required [66]. A list of some commercialized materials is given in Table 8.1.

Table 8.1: Selected commercially available bio-based smart coatings and coating components.

Products	Characteristics	Applications	Producers
Naturacoat	Bio-based anti-graffiti paint	Bio-based sacrificial protective coating	Beardow Adams www.beardowadams.com
Sigma Air Pure	Renewable natural oil-based protective coating	Protection and indoor air purification	PPG www.ppg.com
UCECOAT®7999	Clear coatings aimed for industrial wood applications	High mechanical and chemical resistance	allnex allnex.com/en
VINNECO®	Bio-based acetic acid derivative	Superhydrophobic coatings	Wacker www.wacker.com
Furalkamine Green epoxy-curing agent	Mannich-based hardener derived from pentosane-rich biomass	Specially made for marine coatings	Chemical Process Services Ltd. (Bitrez) www.bitrez.com

8.6 Future scope

Anticorrosion coatings made from natural ingredients and their derivatives with a variety of smart properties, such as self-healing, self-cleaning, antismudge, anti-icing, fire-retardant, antimicrobial, and so on, are likely to be popular topics for future research. Among these coatings, only self-healing and self-cleaning coatings have been thoroughly researched. Other innovative properties of phytochemicals or extracts are yet to be discovered or explored. Nature is an extensive repository of molecules with enormous untapped potential. Future studies should focus on finding and isolating natural compounds and their synthetic derivatives with this multifunctional property. Corrosion maintenance costs a lot of money; hence, this discovery will have a big impact on a country's GDP and development. One of the most demanding future inventions is a smart coating with several smart properties present, in addition to corrosion protection. Also, we are living in an era of several pandemics; the use of sanitizers and disinfectants is not a permanent solution for this problem. The invention of a smart antimicrobial coating is a necessity of the time, to be used in public places and on high-touch surfaces.

8.7 Conclusion

Bio-based smart coatings is a hot topic with a lot of unexplored possibilities. Smart or intelligent coatings need critical consideration, with a wide range of applications, due to the multifunctionality that can be achieved. The list of coatings with smart properties is steadily growing, primarily because dual to multiple functionalities are now possible in most areas owing to natural products that enhance their potential. This chapter is not a comprehensive review on the subject; instead, it emphasizes some of the tasks that these coatings can accomplish in addition to corrosion inhibition, with industrial relevance. Recent developments in smart functionalities with inherent corrosion inhibition and the active constituents are discussed. Plant oils and their derivatives, carbohydrate-based polymers, structural polymers, heterocyclic and open-chain compounds with heteroatom functionalities, etc., show excellent performance. Isolated phytochemicals, as well as their synthetic modifications, hold a ton of potential. Derivatization can improve the potential to a great extent. Nature still has a library of undiscovered and unexplored molecules with a wide range of properties.

One of the most important techniques in recent developments on smart coatings, primarily for anticorrosive applications, is self-healing coatings using self-healing polymers, UV curing, or reservoirs with healing agents. The triggers can release the healing agent, resulting in a long-term/durable healing effect. Polymers' self-healing properties are based on chemical interactions, such as H-bonding and reversible bonds (such as Diel's Alder reaction). Excellent superhydrophobicity with exceptional repelling ability to various corrosive particles, which is maintained regardless of the metal substrate, is the leading technique for self-cleaning coatings. Aside from the self-cleaning and water resistance features, superhydrophobic surfaces have antifouling, anti-icing, antiscratching, and antibacterial properties, all of which help to prevent corrosion.

Within the industry, product development is ongoing, with an emphasis on improved performance, not just to meet client expectations but also to meet regulatory standards. This may increase the production cost. The development of smart coatings, from basic academic research concepts to a profitable industrial product, remains a major concern. To remove the gap between scientific and technological advancements and future market needs, academics and industry scientists will need to put in a lot more work, and time, and need to collaborate. If we succeed, effective multifunctional smart coatings in real-time service will significantly outstrip market expectations without a great increase in market cost. There is a significant opportunity to conduct substantial research in this area. Smart coatings will help in reducing manpower consumption and will improve energy efficiency.

References

[1]   Ghosh SK. Self-healing materials: Fundamentals, design strategies, and applications. Self-Healing Mater. Wiley-VCH Verlag GmbH & Co. KGaA: Weinheim, Germany, 2008, 1–28. https://doi.org/10.1002/9783527625376.ch1

[2]   Zheludkevich M. Self-healing anticorrosion coatings. Self-Healing Mater. Wiley-VCH Verlag GmbH & Co. KGaA: Weinheim, Germany, 2008, 101–139. https://doi.org/10.1002/9783527625376.ch4

[3]   Twite RL, Bierwagen GP. Review of alternatives to chromate for corrosion protection of aluminum aerospace alloys. Progress in Organic Coatings, 1998; 33: 91–100. https://doi.org/10.1016/S0300-9440(98)00015-0

[4]   Kaufmann B, Christen P. Recent extraction techniques for natural products: Microwave-assisted extraction and pressurised solvent extraction. Phytochemical Analysis, 2002; 13: 105–113. https://doi.org/10.1002/pca.631

[5]   Ulaeto SB, Pancrecious JK, Rajan TPD, Pai BC. Smart coatings. Noble Met. Oxide Hybrid Nanoparticles. Fundamentals and Applications Micro and Nano Technologies, Woodhead publishing 2019, Vol. 1, 341–372. https://doi.org/10.1016/B978-0-12-814134-2.00017-6

[6]   Baghdachi J. Smart Coatings, Smart Coatings II, Chapter 1. American Chemical Society, ACS Publications. 2009, Vol. 1002, 3–24. https://doi.org/10.1021/bk-2009-1002.ch001

[7]   Nagappan S, Moorthy MS, Rao KM, Ha CS. Stimuli-responsive smart polymeric coatings: An overview. Hosseini M, Makhlouf ASH. Eds.. Industrial Applications for Intelligent Polymers and Coatings. Springer International Publishing: Cham, 2016, 27–49. https://doi.org/10.1007/978-3-319-26893-4_2

[8]   de Souza MM, de Sá SC, Zmozinski AV, Peres RS, Ferreira CA. Biomass as the carbon source in intumescent coatings for steel protection against fire. Industrial & Engineering Chemistry Research, 2016; 55: 11961–11969. https://doi.org/10.1021/acs.iecr.6b03537

[9]   Liang H, Feng Y, Lu J, Liu L, Yang Z, Luo Y, Zhang Y, Zhang C. Bio-based cationic waterborne polyurethanes dispersions prepared from different vegetable oils. Industrial Crops and Products, 2018; 122: 448–455. https://doi.org/10.1016/j.indcrop.2018.06.006

[10] Aung MM, Li WJ, Lim HN. Improvement of anticorrosion coating properties in bio-based polymer epoxy acrylate incorporated with nano zinc oxide particles. Industrial & Engineering Chemistry Research, 2020; 5: 1753–1763. https://doi.org/10.1021/acs.iecr.9b05639

[11] Hegde MB, Mohana KNS, Rajitha K, Madhusudhana AM. Reduced graphene oxide-epoxidized linseed oil nanocomposite: A highly efficient bio-based anti-corrosion coating material for mild steel. Progress in Organic Coatings, 2021; 159: 106399. https://doi.org/10.1016/j.porg coat.2021.106399

[12] Patil CK, Rajput SD, Marathe RJ, Kulkarni RD, Phadnis H, Sohn D, Mahulikar PP, Gite VV. Synthesis of bio-based polyurethane coatings from vegetable oil and dicarboxylic acids. Progress in Organic Coatings, 2017; 106: 87–95. https://doi.org/10.1016/j.porgcoat.2016.11.024

[13] Liang B, Kuang S, Huang J, Man L, Yang Z, Yuan T, Synthesis and characterization of novel renewable tung oil-based UV-curable active monomers and bio-based copolymers, Progress in Organic Coatings. 2019; 129: 116–124. https://doi.org/10.1016/j.porgcoat.2019.01.007.

[14] da Silva LRR, Avelino F, Diogenes OBF, Sales VDOF, da Silva KT, Araujo WS, Mazzetto SE, Lomonaco D, Development of BPA-free anticorrosive epoxy coatings from agro industrial waste, Progress in Organic Coatings. 2020; 139: 105449. https://doi.org/10.1016/j.porgcoat.2019.105449.

[15] Qian B, Song Z, Hao L, Wang W, Kong D. Self-healing epoxy coatings based on nanocontainers for corrosion protection of mild steel. Journal of the Electrochemical Society, 2017; 164: C54–C60. https://doi.org/10.1149/2.1251702jes

[16] Ataei S, Hassan A, Azari P, Pingguan-Murphy B, Yahya R, Basirun WJ, Shahabudin N. Electrosprayed PMMA microcapsules containing green soybean oil-based acrylated epoxy and a thiol: A novel resin for smart self-healing coatings. Smart Materials and Structures, 2020; 29: 085037. https://doi.org/10.1088/1361-665X/ab9754

[17] Guin AK, Nayak S, Bhadu MK, Singh V, Rout TK. Development and performance evaluation of corrosion resistance self-healing coating. ISRN Corrosion, 2014; 2014: 979323. https://doi.org/10.1155/2014/979323

[18] Yang Y, Urban MW. Self-healing polymeric materials. Chemical Society Reviews, 2013; 42: 7446. https://doi.org/10.1039/c3cs60109a

[19] Zhang C, Liang H, Liang D, Lin Z, Chen Q, Feng P, Wang Q. Renewable castor-oil-based waterborne polyurethane networks: Simultaneously showing high strength, self-healing, processability and tunable multishape memory. Angewandte Chemie International Edition, 2021; 60: 4289–4299. https://doi.org/10.1002/anie.202014299

[20] Chen Y, Xia C, Shepard Z, Smith N, Rice N, Peterson AM, Sakulich A. Self-healing coatings for steel-reinforced concrete. ACS Sustainable Chemistry & Engineering, 2017; 5: 3955–3962. https://doi.org/10.1021/acssuschemeng.6b03142

[21] Arukalam IO, Ishidi EY, Obasi HC, Madu IO, Ezeani OE, Owen MM. Exploitation of natural gum exudates as green fillers in self-healing corrosion-resistant epoxy coatings. Journal of Polymer Research, 2020; 27: 80. https://doi.org/10.1007/s10965-020-02055-y

[22] Zheludkevich ML, Tedim J, Freire CSR, Fernandes SCM, Kallip S, Lisenkov A, Gandini A, Ferreira MGS. Self-healing protective coatings with "green" chitosan based pre-layer reservoir of corrosion inhibitor. Journal Of Materials Chemistry, 2011; 21: 4805. https://doi.org/10.1039/c1jm10304k

[23] Chaudhari AB, Tatiya PD, Hedaoo RK, Kulkarni RD, Gite VV. Polyurethane prepared from neem oil polyesteramides for self-healing anticorrosive coatings. Industrial & Engineering Chemistry Research, 2013; 52: 10189–10197. https://doi.org/10.1021/ie401237s

[24] Suryanarayana C, Rao KC, Kumar D. Preparation and characterization of microcapsules containing linseed oil and its use in self-healing coatings. Progress in Organic Coatings, 2008; 63: 72–78. https://doi.org/10.1016/j.porgcoat.2008.04.008

[25] Samadzadeh M, Boura SH, Peikari M, Ashrafi A, Kasiriha M. Tung oil: An autonomous repairing agent for self-healing epoxy coatings. Progress in Organic Coatings, 2011; 70: 383–387. https://doi.org/10.1016/j.porgcoat.2010.08.017

[26] Ataei S, Khorasani SN, Neisiany RE. Biofriendly vegetable oil healing agents used for developing self-healing coatings: A review. Progress in Organic Coatings, 2019; 129: 77–95. https://doi.org/10.1016/j.porgcoat.2019.01.012

[27] Ulaeto SB, Pancrecious JK, Rajan TPD, Pai BC. Smart coatings. *Noble Met. Oxide Hybrid Nanoparticles*. Woodhead publishing, 2019, 341–372. https://doi.org/10.1016/B978-0-12-814134-2.00017-6

[28] Zhang M, Feng S, Wang L, Zheng Y. Lotus effect in wetting and self-cleaning. Biotribology, 2016; 5: 31–43. https://doi.org/10.1016/j.biotri.2015.08.002

[29] Dodiuk H, Rios PF, Dotan A, Kenig S. Hydrophobic and self-cleaning coatings. Polymers for Advanced Technologies, 2007; 18: 746–750. https://doi.org/10.1002/pat.957

[30] Cheng Y, Miao D, Kong L, Jiang J, Guo Z. Preparation and performance test of the super-hydrophobic polyurethane coating based on waste cooking oil. Coatings, 2019; 9: 861. https://doi.org/10.3390/coatings9120861

[31] Atta AM, Ahmed MA, Al-Lohedan HA, El-Faham A. Multi-functional cardanol triazine Schiff base polyimine additives for self-healing and super-hydrophobic epoxy of steel coating. Coatings, 2020; 10: 327. https://doi.org/10.3390/coatings10040327

[32] Wahby MH, Atta AM, Moustafa YM, Ezzat AO, Hashem AI. Hydrophobic and superhydrophobic bio-based nano-magnetic epoxy composites as organic coating of steel. Coatings, 2020; 10: 1201. https://doi.org/10.3390/coatings10121201

[33] Xie W-Y, Wang F, Xu C, Song F, Wang X-L, Wang Y-Z. A superhydrophobic and self-cleaning photoluminescent protein film with high weatherability. Chemical Engineering Journal, 2017; 326: 436–442. https://doi.org/10.1016/j.cej.2017.05.170

[34] Zheng S, Bellido-Aguilar DA, Huang Y, Zeng X, Zhang Q, Chen Z. Mechanically robust hydrophobic bio-based epoxy coatings for anti-corrosion application. Surface and Coatings Technology, 2019; 363: 43–50. https://doi.org/10.1016/j.surfcoat.2019.02.020

[35] Zhong X, Lv L, Hu H, Jiang X, Fu H. Bio-based coatings with liquid repellency for various applications. Chemical Engineering Journal, 2020; 382: 123042. https://doi.org/10.1016/j.cej.2019.123042

[36] Liang B, Chen J, Guo X, Yang Z, Yuan T. Bio-based organic-inorganic hybrid UV-curable hydrophobic coating prepared from epoxidized vegetable oils. Industrial Crops and Products, 2021; 163: 113331. https://doi.org/10.1016/j.indcrop.2021.113331

[37] Nurioglu AG, Esteves AC, Gijsbertus dW. Non-toxic, non-biocide-release antifouling coatings based on molecular structure design for marine applications. Journal of Materials Chemistry B, 2015; 3(32): 6547–6570.

[38] Kyei SK, Darko G, Akaranta O. Chemistry and application of emerging ecofriendly antifouling paints: A review. Journal of Coatings Technology and Research, 2020; 17: 315–332. https://doi.org/10.1007/s11998-019-00294-3

[39] Armstrong E, Boyd KG, Burgess JG. Prevention of marine biofouling using natural compounds from marine organisms. Biotechnology Annual Review, 2000; 6: 221–241. https://doi.org/10.1016/S1387-2656(00)06024-5

[40] Pan J, Xie Q, Chiang H, Peng Q, Qian P-Y, Ma C, Zhang G. "From the nature for the nature": An eco-friendly antifouling coating consisting of poly(lactic acid)-based polyurethane and

natural antifoulant. ACS Sustainable Chemistry & Engineering, 2020; 8: 1671–1678. https://doi.org/10.1021/acssuschemeng.9b06917

[41] Somisetti V, Narayan R, Kothapalli RVSN. Multifunctional polyurethane coatings derived from phosphated cardanol and undecylenic acid based polyols. Progress in Organic Coatings, 2019; 134: 91–102. https://doi.org/https://doi.org/10.1016/j.porgcoat.2019.04.077

[42] Chen Y, Zhang G, Zhang G, Ma C. Rapid curing and self-stratifying lacquer coating with antifouling and anticorrosive properties. Chemical Engineering Journal, 2021; 421: 129755. https://doi.org/10.1016/j.cej.2021.129755

[43] Li X, Huang T, Heath DE, O'Brien-Simpson NM, O'Connor AJ. Antimicrobial nanoparticle coatings for medical implants: Design challenges and prospects. Biointerphases, 2020; 15: 060801. https://doi.org/10.1116/6.0000625

[44] Kumar S, Ye F, Dobretsov S, Dutta J. Chitosan nanocomposite coatings for food, paints, and water treatment applications. Applied Sciences, 2019; 9: 2409. https://doi.org/10.3390/app9122409

[45] Liang H, Lu Q, Liu M, Ou R, Wang Q, Quirino RL, Luo Y, Zhang C. UV absorption, anticorrosion, and long-term antibacterial performance of vegetable oil based cationic waterborne polyurethanes enabled by amino acids. Chemical Engineering Journal, 2021; 421: 127774. https://doi.org/10.1016/j.cej.2020.127774

[46] Ulaeto SB, Nair AV, Pancrecious JK, Karun AS, Mathew GM, Rajan TPD, Pai BC. Smart nanocontainer-based anticorrosive bio-coatings: Evaluation of quercetin for corrosion protection of aluminium alloys. Progress in Organic Coatings, 2019; 136: 105276. https://doi.org/10.1016/j.porgcoat.2019.105276

[47] Bourbigot S, Le Bras M, Duquesne S, Rochery M. Recent advances for intumescent polymers. Macromolecular Materials and Engineering, 2004; 289: 499–511. https://doi.org/10.1002/mame.200400007

[48] Liang S, Neisius NM, Gaan S. Recent developments in flame retardant polymeric coatings. Progress in Organic Coatings, 2013; 76: 1642–1665. https://doi.org/10.1016/j.porgcoat.2013.07.014

[49] Ma H-X, Li -J-J, Qiu -J-J, Liu Y, Liu C-M. Renewable cardanol-based star-shaped prepolymer containing a phosphazene core as a potential biobased green fire-retardant coating. ACS Sustainable Chemistry & Engineering, 2017; 5: 350–359. https://doi.org/10.1021/acssusche meng.6b01714

[50] He S, Gao -Y-Y, Zhao Z-Y, Huang S-C, Chen Z-X, Deng C, Wang Y-Z. Fully bio-based phytic acid–basic amino acid salt for flame-retardant polypropylene. ACS Applied Polymer Materials, 2021; 3: 1488–1498. https://doi.org/10.1021/acsapm.0c01356

[51] Gao -Y-Y, Deng C, Du -Y-Y, Huang S-C, Wang Y-Z. A novel bio-based flame retardant for polypropylene from phytic acid. Polymer Degradation and Stability, 2019; 161: 298–308. https://doi.org/10.1016/j.polymdegradstab.2019.02.005

[52] Zhang J, Mi X, Chen S, Xu Z, Zhang D, Miao M, Wang J. A bio-based hyperbranched flame retardant for epoxy resins. Chemical Engineering Journal, 2020; 381: 122719. https://doi.org/10.1016/j.cej.2019.122719

[53] Ding H, Wang J, Wang C, Chu F. Synthesis of a novel phosphorus and nitrogen-containing bio-based polyols and its application in flame retardant polyurethane sealant. Polymer Degradation and Stability, 2016; 124: 43–50. https://doi.org/10.1016/j.polymdegradstab.2015.12.006

[54] Xing Y, Li Y, Lin Z, Ma X, Qu H, Fan R. Synthesis and characterization of bio-based intumescent flame retardant and its application in polyurethane. Fire and Materials, 2020; 44: 814–824. https://doi.org/10.1002/fam.2877

[55] Bayer IS. On the durability and wear resistance of transparent superhydrophobic coatings. Coatings, 2017; 7: 12. https://doi.org/10.3390/coatings7010012
[56] https://www.beardowadams.com/products-and-solutions/brands/naturacoat
[57] Liang Y, Zhang D, Zhou M, Xia Y, Chen X, Oliver S, Shi S, Lei L. Bio-based omniphobic polyurethane coating providing anti-smudge and anti-corrosion protection. Progress in Organic Coatings, 2020; 148: 105844. https://doi.org/10.1016/j.porgcoat.2020.105844
[58] Zheng S, Bellido-Aguilar DA, Wu X, Zhan X, Huang Y, Zeng X, Zhang Q, Chen Z. Durable waterborne hydrophobic bio-epoxy coating with improved anti-icing and self-cleaning performance. ACS Sustainable Chemistry & Engineering, 2019; 7: 641–649. https://doi.org/10.1021/acssuschemeng.8b04203
[59] Bellido-Aguilar DA, Zheng S, Huang Y, Zeng X, Zhang Q, Chen Z. Solvent-free synthesis and hydrophobization of biobased epoxy coatings for anti-icing and anticorrosion applications. ACS Sustainable Chemistry & Engineering, 2019; 7: 19131–19141. https://doi.org/10.1021/acssuschemeng.9b05091
[60] Lv J, Zhu C, Qiu H, Zhang J, Gu C, Feng J. Robust icephobic epoxy coating using maleic anhydride as a crosslinking agent. Progress in Organic Coatings, 2020; 142: 105561. https://doi.org/10.1016/j.porgcoat.2020.105561
[61] Raja PB, Sethuraman MG. Natural products as corrosion inhibitor for metals in corrosive media – A review. Materials Letters, 2008; 62: 113–116. https://doi.org/https://doi.org/10.1016/j.matlet.2007.04.079
[62] Dahmani K, Galai M, Ouakki M, Cherkaoui M, Touir R, Erkan S, Kaya S, El Ibrahimi B. Quantum chemical and molecular dynamic simulation studies for the identification of the extracted cinnamon essential oil constituent responsible for copper corrosion inhibition in acidified 3.0 wt% NaCl medium. Inorganic Chemistry Communications, 2021; 124: 108409. https://doi.org/10.1016/j.inoche.2020.108409
[63] Lai X, Hu J, Ruan T, Zhou J, Qu J. Chitosan derivative corrosion inhibitor for aluminum alloy in sodium chloride solution: A green organic/inorganic hybrid. Carbohydrate Polymers, 2021; 265: 118074. https://doi.org/10.1016/j.carbpol.2021.118074
[64] Oguzie EE, Adindu CB, Enenebeaku CK, Ogukwe CE, Chidiebere MA, Oguzie KL. Natural products for materials protection: Mechanism of corrosion inhibition of mild steel by acid extracts of piper guineense. Journal of Physical Chemistry C, 2012; 116: 13603–13615. https://doi.org/10.1021/jp300791s
[65] Latthe SS, Sutar RS, Kodag VS, Bhosale AK, Kumar AM, Kumar Sadasivuni K, Xing R, Liu S. Self – Cleaning superhydrophobic coatings: Potential industrial applications. Progress in Organic Coatings, 2019; 128: 52–58. https://doi.org/10.1016/j.porgcoat.2018.12.008
[66] https://www.futuremarketsinc.com/nanocoatings-in-the-automotive-industry/.

Ruby Aslam, Mohammad Mobin, Jeenat Aslam

9 Commercialization of environmentally sustainable corrosion inhibitors

Abstract: The existing industrial and social world is in a trillion-dollar deficit owing to corroding industrial equipment, bridges, and railway lines. While the world is already incurring a loss due to corrosion, it is critical to evaluate the cost of preventive approaches. Synthetic inhibitors are highly effective corrosion inhibitors, even though they are expensive and hazardous to cater to severe environmental regulations in many countries. The demand for sustainable corrosion inhibitors is rapidly growing because of government regulations imposing severe environmental constraints on the oil and gas producing industry, which require less environmental consequence to be developed and implemented. Environmentally sustainable corrosion inhibitors such as plant extracts, amino acids, biopolymers, and biosurfactants, on the other hand, have limited corrosion inhibition efficiency, according to various studies. Despite the vast number of patents relating to corrosion preventive actions and corrosion inhibitor uses, only a few patents specifically deal with green corrosion inhibitors. As a result, the commercial and economic implications of using environmentally sustainable corrosion inhibitors must be examined.

Keywords: Corrosion, commercialization, economy, GDP, inhibitors

9.1 Introduction: corrosion and its impact

Corrosion of metallic materials is a major issue in material science that has existed since the discovery of metals. Corrosion is the gradual deterioration of a metallic material's qualities because of electrochemical/chemical reactions with its surroundings [1]. Wear, galling, and erosion are three types of physical metal degradation. This process involves introducing a secondary energy feed to achieve the desired metallic form [2]. When subjected to water, liquid chemicals, salts (NaCl), acids (HNO$_3$, H$_2$SO$_4$, HCl, etc.), bases (NaHCO$_3$, CaCO$_3$, NaOH, etc.), corrosive metal polishes and gases (sulfur-containing gases, ammonia, and formaldehyde), metals eventually decompose. Nonmetals like granite erode, plastic surfaces swell or crack, Portland cement leaches away, and wood decomposes or cleaves, whereas "corrosion" is now limited to chemical push over metals.

Acknowledgments: Ruby Aslam gratefully acknowledges the Council of Scientific and Industrial Research (CSIR), New Delhi, India, for the Research Associate fellowship (File Number 09/112(0616)2K19 EMR-I). Thanks are also due to University Sophisticated Instrument Facility (USIF), A.M.U. Aligarh, India, for providing TEM, SEM, and EDS facilities.

https://doi.org/10.1515/9783110760583-009

Boiler tanks, pressure basins, turbines/motor blades, hazardous chemical containers, bridges, automotive routing devices, and airplane parts are all susceptible to corrosion. Radioactive power facilities must pay close attention to safety throughout the construction of equipment and the removal of nuclear waste. Corrosion is ineffective not only with metals but also with water and energy. The importance of corrosion studies is divided into three categories: (i) economic issues, including direct and indirect losses; (ii) safety, including better safety of operational apparatus/tools; and (iii) metal and materials resource protection. Corrosion costs can be classified into three groups, as indicated in Figure 9.1.

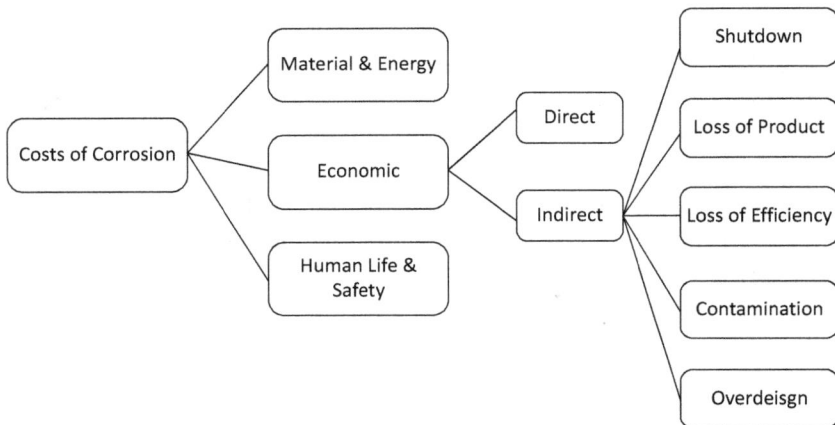

Figure 9.1: Corrosion costs breakdown.

Corrosion scientists and material engineers are working to reduce the corrosive impact on reservoirs, pipelines, coastal structures, machinery components, ships, bridges, and other structural materials, resulting in lower economic losses, improved manufacturing security, and material conservation. In industries, corrosion inhibitors are a cost-effective alternative to conventional corrosion-prevention techniques such as cathodic protection, anodic protection, protective coatings, and material selection [3–6]. If it is achieved through the addition of chemical compounds, any corrosion slowing strategy might be deemed "corrosion inhibition." Corrosion inhibitors are chemical chemicals or additives, which, when used in tiny amounts in a harsh or corrosive environment, minimize or prevent corrosion between metal and media surfaces [7, 8].

Corrosion inhibitors work in two ways: they spread inhibitor molecules over a metal surface; then, their functional groups interact with the metal surface. Inhibitor compatibility in corroding systems, inhibitor solubility in corrosive medium, inhibitor stability when temperature and pH are changed, inhibitor cost, corrosion inhibition efficacy, and eco-friendliness are all important characteristics to consider.

The rate of corrosion is generally higher in an acidic environment when hydrogen gas is liberated through a cathodic reaction. The hydrogen discharged from the metal's surface either recombines and exits as gas or enters the metal, causing hydrogen–activated metal degradation [9]. Corrosion inhibitors are important because they reduce hydrogen evolution through the adsorption process, which inhibits corrosion and protects metal surfaces [10]. Concentration, temperature, pH, solution corrosives, and dissolved salts all affect inhibitor efficacy. Organic compounds, inorganic complexes, rare earth elements, and natural products have all been successfully identified as corrosion inhibitors for various metals in a variety of harsh environments, including aqueous and nonaqueous, molten substances, and dry media. Depending on the medium, metals, and environmental conditions, all these inhibitors were shown to be successful and exclusive in their activity.

Environment-friendly corrosion inhibitors have proven to be an excellent alternative for replacing traditional organic corrosion inhibitors that are costly, harmful, and manufactured. Numerous elements in environmentally acceptable corrosion inhibitors can absorb and prevent metallic corrosion. Some corrosion inhibitors, such as polymers, fungi, plant extraction, bioingredients, and drugs, have been studied in various basic, acidic, and neutral media to justify their efficacy against mild steel, iron, stainless steel, carbon steel, copper alloy, aluminum alloy, and other metals and alloys [11].

This chapter delves into the economic and commercial aspects of using environment-friendly corrosion inhibitors to combat corrosion.

9.2 Global corrosion inhibitor market

Inhibitors of global corrosion: Environmental laws, particularly those governing its disposal and interaction with humans, have a significant impact on market volume. For example, increasing product treatment removes hazardous raw components like chromium (Cr), zinc (Zn), and phosphorus (P) from the production process (P). By 2023, this is predicted to have slowed global industrial growth [12]. Even though manufacturers are promoting environment-friendly products, new production outlooks will emerge. The market is expected to grow at a compound annual growth rate of 4.9% from 2021 to 2026 [13]. The market for corrosion inhibitors is being driven by increased demand from a range of end-use categories, as well as tight regulatory and sustainability consents affecting the environment.

9.2.1 Marketplace dynamics of corrosion inhibitors

9.2.1.1 Driver: decreasing costs of corrosion

Corrosion-related costs incurred by end-use industries are a primary factor influencing corrosion inhibitor consumption. Corrosion costs vary depending on the industry. The yearly corrosion cost to the oil and gas industry in the United States is estimated to be USD 27 billion, according to the National Association of Corrosion Engineers (NACE) International. The expenses of corrosion can be reduced by using more corrosion-related technical procedures and using corrosion-resistant materials. Metal corrosion is prevented using corrosion inhibitors. They act by forming an absorbent coating or delaying the onset of anodic and cathodic processes. These inhibitors are used in various industries to reduce protection and maintenance costs, extend the useful life of tools, and reduce industrial losses due to corrosion. This lowers the cost of corrosion and stimulates the market for corrosion inhibitors.

The increasing demand from various end-user industries such as mining and oil and gas in emerging economies, such as China, India, Malaysia, Vietnam, Brazil, Chile, and South Africa, is also expected to boost the growth of the global corrosion inhibitor market. In addition, during the forecast period, the increasing number of new oil fields and the construction of new power plants are expected to fuel the market's expansion in these countries.

9.2.1.2 Restraint: use of corrosion-resistant materials

Due to the increasing environmental concerns and the need for more sustainable solutions, the demand for corrosion inhibitors is becoming more prevalent in various industries. The use of exotic materials such as super duplex steel (SDSS) and duplex stainless steel (DSS) is becoming more prevalent in the oil and gas industry due to their high corrosion resistance. Exotic metals, in addition to corrosion resistance, have exceptional durability, strength, and the capacity to withstand extreme heat and pressure, among other attributes. These metals are used in difficult industries such as aircraft, automobiles, oil and gas, and power generation because of their properties.

9.2.1.3 Prospect: growing industrialization in rising economies

China, India, South Korea, and Brazil's robust growth are expected to boost their industrial activities and increase their consumption of corrosion inhibitors. According to Statista [14], the countries that spent the most on infrastructure development and protection were China, and Australia. Other countries such as the United States, the UK, Australia, and the UK also spent less than 1% of their GDP on infrastructure in

2018. In 2020, China has planned to spend about USD 1.07 trillion on building and protecting the country's infrastructure [15].

According to the World Bank, the report noted that the growing economies must spend 4.5% of their gross domestic product on infrastructure development to achieve long-term growth. Over the next five years, the demand for corrosion inhibitor products is expected to increase due to the infrastructure projects that are connected to energy and water needs. According to Oxford Economics, the total amount of money needed to build infrastructure globally is expected to reach USD 94 trillion by 2040 [16]. According to a report by Oxford Economics, the rapid growth of China, India, and Indonesia, the Philippines, Malaysia, Vietnam, and Thailand will create more than half of the global infrastructure costs by 2040 [16]. The rising cost of industrial water in emerging economies will allow manufacturers to offer a wider variety of corrosion inhibitors to protect metal.

9.2.1.4 Challenges: need for sustainable corrosion inhibitor

Strict, long-term legislation is one of the most important factors limiting the growth of the corrosion inhibitor industry. Corrosion inhibitor manufacturers have been subjected to tight regulations due to rising concerns about the chemical's influence on the environment and human health. Currently, corrosion inhibitor manufacturers are encouraged to choose nonhazardous solutions. The ability of nonhazardous corrosion inhibitor formulations to function in hostile environments makes it difficult for companies to provide a substitute for standard formulations. Zinc phosphate, for example, has been used as a substitute in the oil and gas business. This chemical is not as effective as a chromate complex. The US Department of Labor stated that it causes cancer among humans and animals.

These days, biodegradability, sustainable development, cleaner production, and pollution protection are to be employed in manufacturing unit operations, preparations, and applications. In general, regular corrosion inhibitors are risky for the surroundings. This activated the requirement of environment-friendly corrosion inhibitors. These corrosion protective materials are, most of the time, derived from natural resources. Thus, these are far less dangerous to the atmosphere upon negative response. These inhibitors are also termed sustainable corrosion inhibitors, which, via a small dose, can successfully decrease the corrosion problem on the surface of metals and alloys. Considering ecological concerns, sustainable corrosion inhibitors can be utilized in place of conventional inhibitors, yet the consequences of the sustainable inhibitors on the atmosphere must be taken into consideration. PARCOM (Paris Commission) delegated via EEC (The European Economy Community) gives a few guidelines, which are, briefly, as follows. PARCOM guidelines are based on three parameters:

(a) **Toxicity** is determined as LC50 (Lethal Concentration that influences the population by 50%) as well as EC50 (Effective Concentration of chemicals mandatory for which the population is influenced via 50%). In general, LC50 is higher than EC50.
(b) **Biodegradation** is an assessment of the time during which a chemical perseveres in the surrounding area of the system. This analysis is followed by the Organization for Economic Co-operation and Development (OECD) 301D examination observing 28 days of biodegradation study.
(c) **Bioaccumulation** is determined by calculating the division co-efficient of a chemical dispersed among two phases of water and 1-octanol. The chemical compounds present in inhibitors are investigated via the HPLC OECD 117 process. A model proposed by PARCOM, which is termed CHARM (Chemical Hazard Assessment and Risk Management), includes four modules: hazard assessment, prescreening, risk management, and risk analysis. Hazard assessment is completed by calculating the ratio of PEC (predicted environmental concentration) to NEC (no effect concentration). The environment system will not be affected if NEC is equivalent or better than PEC. Pre-screening applies the term "bad actors," which specifies the biodegradation level is less than 20% or the logarithmic value of the partition coefficient is bigger than 0.5, and the molecular weight is also larger than 700. Risk management recognizes the steps to be taken to decrease the destructive effects on the atmosphere involving unconventional chemicals, cost-effectiveness, and treatment process. It brings forth two terms: best environmental practice (BEP) and best available technology (BAT). Risk analysis determines the probability of occurrence of destructive effects of the chemicals on the atmosphere.

9.2.2 Market of corrosion inhibitors by product

9.2.2.1 Environment-friendly corrosion inhibitors

Environment-friendly corrosion inhibitors are predictable to witness the maximum development, with a CAGR of 7.0%, by 2026, because the novelty majorly focuses on the utilization of green and environment-friendly compounds, for example, plant extracts containing various organic compounds.

9.2.2.2 Water-based corrosion inhibitors

The marketplace volume for water-based corrosion inhibitors in 2018 is expected to have a superior development, with a CAGR of 4.7%, because of the rising need for water treatment chemicals owing to its lesser volatile organic content (VOC) emissions.

9.2.2.3 Organic corrosion inhibitors

The marketplace dimension for organic corrosion inhibitors controls internationally in 2018 is estimated to have a maximum expansion, with a CAGR of 4.5% during the estimated time, mostly owing to its better features including stability, and broad practice in industrialized applications.

9.2.2.4 Inorganic corrosion inhibitors

Inorganic corrosion inhibitors marketplace volume is expected to witness moderate gains of more than 3% CAGR over the expected timeframe. Inorganic manufacturing can be isolated as cathodic and anodic products based on its action mechanism on metals, for example, magnesium (Mg), zinc (Zn), and nickel (Ni). The product is broadly employed due to its superior stability in cruel climatic situations.

9.2.3 The market of environment-friendly corrosion inhibitors, by application

In 2019, the global corrosion inhibitors market was valued at USD 7.4 billion, and it is predicted to increase at a CAGR of 3.8% from 2020 to 2027. The expansion of bio-based and environment-friendly corrosion inhibitors can be attributed to the market's development. A constant movement toward the development of environmentally benign corrosion inhibitors could help the industry grow sustainably and increase corrosion inhibitor adoption among long-term end users. A corrosion inhibitor works by forming a barrier over the covered surface to keep moisture out, which is the primary cause of erosion. Cathodic activity, anodic activity, interfacial activity, adsorption activity, and barrier activity are some of the methods that organic compounds use.

9.2.3.1 The end-use market of environmentally-friendly corrosion inhibitors

In 2019, the organic sector had the largest market share of 73.8%. Environment-friendly corrosion inhibitors have made significant development in recent years. These organic substances are made from plant extracts such as spices and aromatic herbs, and they are cheaper and nontoxic than toxic. The aromatic structures of these compounds can be found in chains and free electron pairs. For example, rosemary leaves and Delonix regia stop aluminum from corroding, whereas natural honey keeps copper from corroding. The qualities are predicted to promote section development throughout time.

In 2019, the oil and gas sector had the largest market share of 33.1%. Internal corrosion concerns in refineries, pipelines, and petrochemical plants are becoming more common. Due to rapid industrialization and urbanization, the market for environment-friendly corrosion inhibitors is gaining traction in the power generating sector in Asia Pacific's rising economies. Erosion has been the most significant concern in power facilities, resulting in critical downtime. As the metal components are constantly in touch with water, the steam circuits in nuclear, thermal, and hydroelectric plants are prone to degradation. Most power plants check characteristics such as conductivity, pH, and the presence of corroding cations and anions, which indicate the rate of corrosion, to assure effective power generation. Furthermore, they confirm that corrosion inhibitors such as phosphate, phosphonates, triazoles (for copper), and zinc are present in suitable quantities (for steel). Furthermore, paper and pulp manufacturing operations are exposed to extremely corrosive environments. Water, air, and organic contaminants are then exposed to instruments including recovery boilers, digesters, evaporators, bleachers, storage tanks, and paper manufacturing equipment. Bleaching is a process used to whiten and brighten paper and to produce wastewater. It produces a variety of chlorinated compounds, which can cause significant damage to the materials in contact with them.

9.2.3.2 Market of corrosion inhibitors by region

In 2019, Asia Pacific had the largest market share (36.3%) and is predicted to be the fastest growing market, soon. The market's growth can be attributed to the region's rapid industrialization, which has sparked demand from power generation and a variety of other end-use sectors. Metalworking and chemical divisions in emerging economies in the region, including China and India, are likely to make a significant contribution to the market soon.

Due to the massive expenditure of water for industrial segments such as ethanol and sugar, as well as petrochemical industrialization, Europe is a well-known user of corrosion inhibitors. Water treatment difficulties in Europe play a critical role in corrosion control activities, as well as maintaining the operational integrity of heat transfer instruments and reducing the negative influence on the energy efficiency of process units.

9.2.3.3 Competitive market share

The global environment-friendly corrosion inhibitors marketplace volume was reasonably consolidated with main industry players including Ecolab, BASF, GE, and Ashland comprising less than half of the total share. A few other important industry share contributors comprise Dow, DuPont, Champion Technologies, AkzoNobel, Lubrizol,

Daubert, Cytec, Eastman, Henkel, Dai-Ichi, Cortec, and Halox. These industry players are widely involved in planned development activities for example capacity acquisitions, expansion and mergers, and novel product growth to enlarge their global presence.

9.2.4 Recent patents on corrosion inhibitors

Corrosion inhibitors are useful as sole inhibitors or as additives in chemical formulations. They are used in cooling water systems, reinforced concrete, oil and gas fields, acid pickling, paint pigments, steam condensate lines, lubricants, and military equipment, among other things. This section looks at several recent patents (from 2011 through 2021) that deal with the use of corrosion inhibitors in a wide range of applications:

Matulewicz et al. [17] were able to create a corrosion inhibitor compound that combines various triazoles. They used a mixture of tolyl triazole, benzotriazole, and toluyl triazoles to prevent the corrosion of the copper and rich copper alloys after exposure to 10 ppm sodium hypochlorite. When compared to a corrosion inhibitor component that is 100% of the other triazoles, the composition reduces the overall corrosion rate by at least 0.05 mpy. Malwitz et al. [18] filed a patent for compositions and methods for ecologically acceptable corrosion inhibitors for the oil and gas industry. The invention is more precisely directed to formulations including the general formula (Figure 9.2) for a quaternary nitrogen-containing corrosion inhibitor.

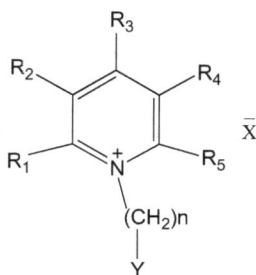

Figure 9.2: The general formula given for the used inhibitor.

The authors choose mild steels, such as N80, L80, and J55, as well as high chromium steels, such as 13Cr85. They also consider various kinds of acids and anhydrides as aggressive media.

A patent was submitted by Garima Misra and Arunesh Kumar [19] for the treatment fluid, which consists of a carboxylate, a water-based hydric acid, and a corrosion inhibitor that is made up of amines. For an experiment at 200 °F, the treatment fluid lost less than 0.05 lb per ft^2. The same fluid without the corrosion inhibitor had a corrosion-resistant weight loss of over 0.05 lb per ft^2.

Schacht et al. patented a vaporizing cleaner that can effectively remove stains and corrosion from various metals, such as steel [20]. The same fluid without the inhibitor had a weight loss of more than 0.05 pounds per ft². Obot and colleagues [21] suggested a corrosion inhibitor with concentration of 0.1–10% of gelatin in the solution and a concentration of 10–28 wt% in the acidic treatment fluid. In 15% HC1 at 25 °C, the gelatin has high inhibitory efficiency for steel, with inhibition rates ranging from 55% to 85% and metal corrosion rates ranging from 12 to 36 mpy. Obot et al. [22] also patented corrosion-inhibiting compositions that included a substituted benzimidazole, a mercapto carboxylic acid, a 2-thioxodihydropyrimidine-dione, sulfhydryl alcohol, a surfactant, and a solvent for carbon steel. In the petroleum industry, the formulations were efficient against corrosion of metallic substrates in sweet (CO_2), sour (H_2S), and/or high salinity conditions. The compositions had inhibitory efficiency of 30–98% and a corrosion rate of 0.4–9 mpy on the metallic substrate. In another patent, Obot et al. [23] proposed a methodology for preventing metal corrosion while treating an oil and gas well. The method involves treating an oil and gas well with an acidic treatment fluid that includes a mixture of 10–28 wt% acid and 0.01–5% 2,3 pyrazine dicarboxylic acid, pyrazine-2-carboxamide, 2-methoxy-3-(1-methyl propyl) pyrazine.

Obot et al. [24], further patented a method of preparing pyrimidine derivatives via a multicomponent condensation reaction. These compounds are effective against corrosion of metallic substrates in acidic (i.e., CO_2), and high salinity environments commonly found in the oil and gas industry, in the concentration ranges of 0.1–200 ppm and exhibited 85–98%.

Obot et al. showed [25] how to inhibit metal corrosion using a liquid containing 0.01–5% pyrazine. This chemical can be commonly used as a corrosion inhibitor after treating various metals with acid stimulation. In addition to pyrazine, this treatment fluid can also be made up of other compounds such as 2,3-pyrazine dicarboxylic pyrazine-2-carboxamide, 2-methoxy-3-(1-methyl propyl) pyrazine, or combinations thereof.

Phoenix clactylifera seed was employed by Rexin et al. as a bio-waste corrosion inhibitor for steel [26]. At 50 °C, the inhibition efficiency was around 97%. In another patent, the compound includes the extract of various plant parts, such as the seeds of different plant species, the leaves of different plant types, and/or mixes [27].

A new type of product was created using the stems and leaves of sweet potato and blossom lettuce to create biodegradable products that are more effective at reducing pollution levels [28]. Utilization of fruit skin extract as corrosion inhibitors was established by Gomes et al. [29]. They noted that these substances could be used by removing the skins and seeds from fruits. Another patent describes the process of extracting plant extract for use as an anticorrosion agent for various metals such as copper and aluminum [30]. Another patent aims to involve the use of vegetable oils as corrosion inhibitors to improve the performance of petroleum refineries. The concept of combining renewable resources with the chemical industry to develop biorefining methods is also proposed [31].

9.3 Impact of COVID-19 pandemic on the global market of corrosion inhibitor

A corrosion inhibitor is a substance that is used in industrial applications such as oil and gas production, water treatment, and various other processes. They are primarily used to regulate or slow down the corrosion process in a variety of industries, as well as to treat hazardous water. The chemical industry has been seriously affected around the world because of the present pandemic. Manpower shortages, resource scarcity, logistical constraints, and other factors have all hampered the industry's growth significantly. Though the corrosion inhibitor industry has been relatively unaffected, demand remains strong in areas such as oil and gas, pharmaceuticals, and others.

9.4 Conclusion and future scope

Green chemistry has proven to be a promising research field in the subject of metallic materials deterioration, which is often addressed with hazardous materials. Corrosion inhibitors are known to be commonly used in metal corrosion control. Green chemistry has become an important aspect of modern life to minimize waste and protect the environment. Green corrosion inhibitors made from naturally occurring substances, on the other hand, are rarely utilized to reduce damage in real-world plants. This element is closely linked to the common mistake of using green components from fruits, vegetables, or matrices as inhibitors. They are essential since they are hard to come by, edible, and in small quantities. Despite the number of experimental works that have been done, the question of green inhibitors is still unsolved. The high level of interest in the topic boosted research, resulting in many compounds being evaluated. However, the most often used methods are conventional and thus unsuitable for adequately characterizing potential efficacy of inhibitors. Many of the greener options for extracting organic corrosion inhibitors are edible and valuable for human needs in a multitude of areas, including medicinal, pharmaceutical, and food usage, making them highly useful. There is still a gap between research and industry implementation of these green corrosion inhibitors. According to the investigation, many industries are still using corrosion inhibitors that have been proven to be costly and environmentally unfriendly, even though researchers are working tirelessly to develop cheap, environment-friendly, and readily available green corrosion inhibitors from greener sources materials. Wastes can provide valuable substances at a low cost and with ease of access. The use of these bioactive residues as a platform for producing valuable chemicals is efficient, affordable, and environmentally benign. Future studies should concentrate more on wastes from greener sources, which are an environmental nuisance that harms both humans and aquatic life.

References

[1]   Kelly RG, Scully JR, Shoesmith DW, Buchheit RG. Electrochemical Techniques in corrosion science and engineering. Marcel Dekker, Inc: United States of America, 2003.

[2]   Roberge PR. Corrosion inspection and monitoring. John Wiley Publication: New Jersey, 2007.

[3]   Bregman JI. Corrosion Inhibitors. Collier MacMillan Co: London, 1963.

[4]   Eldredge GG, Warner JC. In Uhlig HH. Ed. *The Corrosion Handbook*. Wiley: New York, 1948.

[5]   Putilova N, Balezin SA, Barannik VP. Metallic Corrosion Inhibitors (translated from Russian). Pergamon Press: London, 1966.

[6]   Ranney MW. Inhibitors – Manufacture and Technology. Noyes Data Corp: New Jersey, 1976.

[7]   Riggs Jr. O. L., Nathan C.C. *"Corrosion Inhibitors (2nd Edition),"* Houston, 1973.

[8]   Papavinasam S. Corrosion Inhibitors. Uhlig's Corrosion Handbook, 2nd Edition. John Wiley Publications: New York, 2000, 1089–1105.

[9]   Sanatkumar BS, Jagganath N, Nityananda SA. Influence of 2-(4-chlorophenyl)-2-oxoethyl benzoate on the hydrogen evolution and corrosion inhibition of 18 Ni 250 grade weld aged maraging steel in 1.0 M sulfuric acid medium. International Journal of Hydrogen Energy, 2012; 37: 9431–9442.

[10] Sanatkumar BS, Jagganath N, Nityananda SA. The corrosion inhibition of maraging steel under weld aged condition by 1(2E)-1-(4-aminophenyl)-3-(2-thienyl)prop-2-en-1-one in 1.5 M hydrochloric acid medium. Journal of Coating and Technology Research, 2012; 9: 483–493.

[11] Verma C, Ebenso EE, Bahadur I, Quraishi MA. An overview on plant extracts as environmentally sustainable and green corrosion inhibitors for metals and alloys in aggressively corrosive media. Journal of Molecular Liquids, 2018; 266: 577–590.

[12] https://www.gminsights.com/request-sample/detail/389

[13] https://www.marketsandmarkets.com/Market-Reports/Study-Corrosion-Inhibitor-Market -246.html

[14] https://www.statista.com/statistics/566787/average-yearly-expenditure-on-economic-infrastructure-as-percent-of-gdp-worldwide-by-country/

[15] https://www.chinabankingnews.com/2020/03/17/china-touts-50-trillion-yuan-in-infrastructure-spending-yet-just-7-5-trillion-scheduled-for-2020/

[16] https://cdn.gihub.org/outlook/live/methodology/Global+Infrastructure+Outlook+-+July +2017.pdf

[17] Matulewicz WN, Vogt P, Milawski J Corrosion inhibitor compositions comprising tetra hydrobenzotriazoles and other triazoles and methods for using same, U.S. Patent 20120308432 A1, 2012.

[18] Malwitz M, Woloch D Environmentally friendly corrosion inhibitor, U.S. Patent 20130112106 A1, 2013.

[19] Misra G, Kumar A Treatment fluid containing a corrosion inhibitor of a polymer including a silicone and amine group, Justia patent 8969263, 2015.

[20] Schacht PF, Schmidt EV Aqueous acid cleaning, corrosion and stain inhibiting compositions in the vapor phase comprising a blend of nitric and sulfuric acid, WO Patent 2012093372 A3, 2013.

[21] Obot IB, Sorour AA, Methods of inhibiting corrosion in acid stimulation operations, US 2020/ 0339871, 2020.

[22] Obot IB, Onyeachu IB Corrosion inhibiting formulations and uses thereof, US010844282B2, 2020.

[23] Obot IB, Umoren SA Methods of inhibiting corrosion with a pyrazine corrosion inhibitor, US010626319B1, 2020.

[24] Obot IB, Onyeachu BI, Quraishi MA, Heterocyclic corrosion inhibitor other publications compounds and uses thereof, US011118272B2, 2021.

[25] Obot IB, Umoren SA Corrosion inhibition method for downhole metal tubing, US 10,988,670 B2, 2020.

[26] Rexin TG, Kumar V Green Corrosion inhibitor for steel in acid medium, 6278/CHE/2014 A. 2014.

[27] Indian Oil Corporation Limited naturally derived corrosion inhibitors composition, process for preparing the same and use thereof 2008. Patent Application, 2008 Mar 10.

[28] Extract corrosion inhibitor of sweet potato stems and lettuce flower stalks and preparation method thereof. No. CN102492948B. Patent, 2013.

[29] Ponciano GJA, Cardoso RJ, D'Elia E Use of fruit skin extracts as corrosion inhibitors and process for producing same. US8926867B2. U.S. Patent, 2015.

[30] Kinlen PJ, Pinheiro SPL Methods and apparatuses for selecting natural product corrosion inhibitors for application to substrates. US 2018/0202051 Al. U.S. Patent, 2018.

[31] Lima R, Casalini A, Palumbo A, Rivici G Corrosion inhibitor comprising complex oligomeric structures derived from vegetable oils. WO 2017/140836 Al. Patent, 2017.

Richika Ganjoo, Shveta Sharma, Humira Assad, Abhinay Thakur,
Ashish Kumar

10 Challenges and future outlook

Abstract: Organic corrosion inhibitors are used in a wide range of industrial applications. However, there are a lot of limitations to using these compounds. The limited
solubility of organic corrosion inhibitors, especially in polar solutions, is among the
most problematic elements of using them. Owing to their water-repellent nature, organic corrosion inhibitors, particularly those comprising aromatic rings and nonpolar
hydrocarbon groups, have restricted solubility, reducing their preventative competence. Corrosion inhibitors, on the other side, have significant practical significance
because they are extensively used during reducing metallic morsel throughout the
production process and minimizing the risk of excessive wear, both of which can result in the abrupt cessation of production methods, and further in considerable economic loss. As a result, we explore the future perspective and obstacles in corrosion
research, such as the drawbacks of various corrosion inhibitors and corrosion inhibition techniques, in this chapter.

Keywords: Corrosion, electrochemical impedance spectroscopy, agriculture waste,
green corrosion inhibitor, weight loss

10.1 Introduction

Corrosion is a major cause of both safety concerns and financial loss. Corrosion scientists and engineers have devised a number of corrosion mitigation measures in
light of the above [1, 2]. Due to the presence of aggressive compounds on metallic
equipment, corrosion is the most significant hazard to businesses and refineries.
Corrosion occurs in many ways depending on how the material interacts with the
environment [3]. Corrosion damage not only results in significant costs for rehabilitation and replacement of diverse equipment but also possess a public safety issue
[4]. As a result, efficient corrosion inhibitors (CI) must be developed. The importance of electron-rich functional groups and p-electrons in the frameworks of CIs is
emphasized in their selection [5]. The majority of well-known inhibitors are organic
compounds with heteroatoms such as N, S, and O, as well as numerous bonds that
enable the compounds to adsorb on steel surfaces. Drugs, plant extracts, ionic
liquids (ILs), surfactants, and other types of substances have lately been employed
as CIs [6]. The goal of this chapter is to highlight the problems that are encountered
when researching inhibitor corrosion protection.

https://doi.org/10.1515/9783110760583-010

10.2 Drawbacks of plant extracts as corrosion inhibitors

It is also vital to use CIs to minimize mineral degradation and to keep acid usage to a minimum [7–10]. CIs are broadly applied in suppressing or at minimum decreasing metal dissolving in a range of disciplines, spanning from economic industries to construction vehicles to surface treatments for cultural artifacts [11]. This is due to their low cost and convenience of usage. Inorganic chemicals, primarily nitrites, chromates, borates, molybdates, silicates, and zinc salts, were used to mitigate corrosion before 1960. These compounds benefit from adsorption by generating a very efficient inactive covering over the metallic surface (passivators). Furthermore, anodic inhibitors were used because they act by stopping the anode activity and assisting the basic inclination of surface modification of the metallic outermost surface, as well as by generating a covering deposited on the metal substrate. Although this model of management is successful, it is hazardous since significant regional invasions can develop if some parts are left unprotected due to inhibitor fatigue. Anodic inhibitors, on the other hand, were commonly used, and a number of them have undesirable features. When such inhibitors are used in extremely tiny amounts, they promote degradation such as pitting, and thus anodic CIs are designated as hazardous. Furthermore, because passivating inorganic inhibitors might induce localized invasion if the passive layer is disintegrated in an acidic medium, organic inhibitors are recommended, especially in acidic conditions. However, between the 1960s and 1980s, they were mostly interchanged by less expensive substitutes including H_3PO_3, surface-active chelates, $C_6H_{12}O_7$, phosphonates/polyphosphonates, $(C_3H_3NaO_2)_n$[polyacrylate], $[NH_4PO_3]_n(OH)_2$, and $(M(RCOO)_n)$.

These compounds are known as precipitating inhibitors (precipitators) because they usually deposit at the metal–environment interface. Nonetheless, environmental concerns came to the fore after that, and between 1980 and 1995, harmful chemicals were supplanted with natural remedies like "natural and bio-polymers, vitamins, bio-surfactants, tannins, and natural compounds. Environment-friendly strategies, like the usage of REM (rare earth metals), poly-functional complexes, the synergism offered by organic compounds or inorganic chemicals using REM, and inhibitor encapsulation," have recently gained a lot of attention. These substitutes have exceptionally low or no toxicity and a greater level of shielding performance. Since the interaction of organic chemicals with "E4 elements encompassing efficiency (proficiency), economy (low-cost), ecology, and environmental friendliness," they are now recognized as one of the most efficient and advantageous strategies of corrosion inhibition [12–14]. Organic CIs are used in a wide range of industrial applications. However, there are a lot of limitations to using these compounds. The limited solubility of organic CIs, especially in polar solutions, is among the most problematic elements of using them. Owing to their water-repellent nature, organic CIs, particularly those comprising arenes and nonpolar hydrocarbon (HC) groups, have restricted dispersibility, reducing their preventative competence. As a result, current corrosion science and engineering investigation is based on developing CIs having hydrophilic polar active/different functional groups in their molecular structures [15]. Besides that, as the popularity for sustainable and green innovations has grown, green marketing has emphasized the importance of environmental sustainability and people's health in a financially feasible fashion over the last couple of years, intending to prevent the use of commercially available destructive and hazardous CIs. Hence, scientific and technological research is focusing on the fabrication, scheme, manufacture, and utilization of ecologically beneficial chemical moieties to interchange conventional dangerous compounds. This is due to the growing demand for conservation knowledge and rigorous environmental regulations. Many renewable, low-cost, widely available, and environmentally compatible alternative solutions to traditional harmful protective coatings stemming from natural reserves, including biopolymers, plant excerpts, and drugs (chemical antibiotics), are now enormously utilized to substitute detrimental CIs. Because of their natural and biological provenance, as well as their nonbioaccumulation property, they are often used as CIs and are considered an ecologically sound substitute for traditional damaging CIs [16]. Moreover, plant extracts' medicinal compounds and phytochemicals (Figure 10.1) work as excellent metallic CIs, owing to their intricate morphologies, which include abundant electron-rich cores, polar functional moieties, and multiple bonds. Because of their biological and natural sources, they are regarded as ecologically responsible substances. However, due to the increasing price of therapeutic agents, their usage has subsequently been curtailed, supporting the concept of employing outdated medicines as ecologically appropriate CIs. Expired drugs, on the other hand, should be examined even more as CIs. Furthermore, considering the current research drug usage from several pharmacotherapeutic moieties as Al CIs, numerous difficulties must be thoroughly clarified before this sector can be effectively accredited. Among the pharmaceuticals evaluated as prospective CIs for Al, antibacterial and antifungal (antimicrobial) therapies are promised candidates.

But there are nevertheless several unsolved queries about the inhibition of corrosion processes of these chemicals, and their chemical interactions with Al should be studied thoroughly. Additionally, as Gökhan Gece [17] has demonstrated, not all drugs are easily biodegradable, and their transition compounds may be similar or even more environmentally damaging. As a result, more research is required before certain CIs may be labeled as green. On the one hand, nanomaterial uses in protection against corrosion are currently being developed. Green inhibitors, on the other hand, provide a far healthier and more ecologically friendly option. As a result, green chemicals are unquestionably the most exciting area of CI exploration. Because of their natural and biological origins, plant extracts are garnering a lot of interest in the modern corrosion inhibition sector, which comprises metal deterioration [17] and corrosion prevention through optimal design, selection of materials, and usage of CI. However, there are a few factors to take into interpretation prior such CIs can be

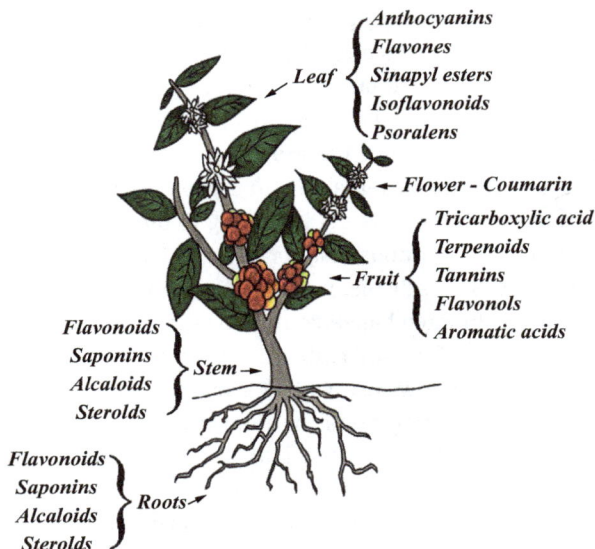

Figure 10.1: Basic parts of a plant and their common active compounds [figure adapted with permission from reference [16]. Copyright The Authors, some rights reserved; exclusive licensee [MDPI]. Distributed under a Creative Commons Attribution (CC BY) License 4.0 https://creativecom mons.org/licenses/by/4.0/].

utilized in real implementations. It is required to determine the "green" value of the natural items under consideration. Toxic effects, biodegradability, and bioaccumulation must all be addressed when evaluating what is and is not green, and this is still a development challenge. The authors of the previous study made no mention of how the plant components were obtained. However, because the excerpts are obtained from nature, there is a high probability that they are safe. Some of them have previously been utilized in other application areas, where their nontoxicity had been demonstrated, at least in part. This also pertains to the pharmaceuticals described, as they've already been medically shown acceptable and are still being used in people's pharmacotherapy nowadays. Unquestionably, the definite quantity of exposure/intake must be taken into account; yet, the latter facts cannot be refuted. Furthermore, the research on inhibition proficiency of medicines for protecting Al and related alloys from decomposition is limited but intriguing, and as such could constitute a challenge for upcoming research. The German WGK classification can be used to swiftly assess environmental acceptability because statistics on this categorization is easily accessible from maximum chemical suppliers. The German Water Hazard Class is abbreviated as WGK (Wassergefährdungsklassen). According to the national German VwVwS regulation (German: Verwaltungsvorschrift wassergefährdende Stoffe), all compounds are either classed as harmless to H_2O or allocated to one out of three water hazard groups, WGK_1, WGK_2, and WGK_3, with cumulative water menace. The bottommost class, WGK_1, is nearly innocuous and nonhazardous [16]. Another method for assessing poisonousness is to use the LC50 or EC50 sets (fatal or effectual intensities in 50% of the verified participants, respectively), which rate substances from extremely toxic to essentially innocuous. Furthermore, biodegradability is often measured after a period of 28 days by the components' 60% perseverance in the atmosphere. Moreover, the partition coefficient Po/w (the partition between 1-octanol and water) is used to assess bioaccumulation. The greater this partition coefficient, the greater the probability of bioaccumulation of the chemical. Typically, publications do not mention the particular constituents found in the excerpts. Before assessing an extract's corrosion inhibition performance, it would be advantageous to separate and analyze individual components, such as "high-performance liquid chromatography (HPLC)-mass spectrometry (MS) or gas chromatography (GC)-MS analysis," to determine the inhibition efficacy of excerpts for which chemicals are accountable. The extraction procedure must be specifically selected. Given the numerous problems related to conventional harvesting treatment plans' high processing temperature changes and long processing times, there is an immediate necessity to encourage the creation and advancement of substitute withdrawal methods that do not entail brutal working settings like elevated operating temperature and hazardous solvents of organic compounds. Furthermore, high-pressure excavation with supercritical fluids is one of the most practical and cost-effective ways for isolating these compounds from plant resources at low temperatures, minimizing rheological features, and eliminating the use of harmful solvents. Water can be used as an alternate solvent or cosolvent to change the polarity of a solvent because it is the cheapest. Subcritical water extraction of phenolic compounds has become an increasingly interesting alternative technique. The dielectric constant of waterfalls and the ionic product improves when the temperature rises above 100 °C. This means that both inorganic and organic components can be extracted using the same solvent. The most significant impediment is the enormous amount of energy required to activate the medium. Additionally, at its pivotal stage (374 °C, 221 bar), water is exceedingly corrosive, which could be an issue for the high-pressure system and infrastructure required to manufacture these items. It would be useful if there was a pressing requirement to support the progress and utilization of alternate extraction processes that did not demand rigorous operating conditions before establishing the performance of an extract's corrosion inhibition, taking into account the numerous issues associated with traditional extraction procedures' greater computational temperatures and long processing times, as well as potentially toxic organic solvents. With a large number of articles on the usage of plants as CIs for various metals especially mild and stainless steel, copper, and aluminum, it is vital to consider the utilization of excerpts for a variety of further metal compounds and alloys, for instance, nickel, zinc, and magnesium. These GCIs (green CIs) still have room to improve their efficacy. Furthermore, more precise research of the inhibitory mechanisms should be addressed employing electrochemical and surface analysis approaches. Other physiochemical factors (e.g., temperature, pH, the strength of ions, and concentration) must be frequently examined before their application becomes more practicable. If authentic corrosion inhibition formulas are implemented, these inhibitors will be considerably more efficient. According to a literature review, several extracts have lately been analyzed to have the potency to hamper the disintegration of materials in corrosive media and other electrolyte mediums. Nonetheless, as there is little information on the corrosion protection properties of plant extracts in NaCl-based electrolytes, the potency of GCIs in these electrolytes must be investigated. Hence, the scope of green inhibitors is anticipated to expand in the future as additional breakthroughs in this field are made. The progress of GCIs and prevention approaches is extremely popular nowadays attributed to the increasing requirement for green chemistry in all disciplines of research and industry. Due to their biological and natural origins, and also their eco-friendly separation, plant excerpts can be considered as cleaner and greener substances to be utilized as inhibitors for metallic substances and their degradation of alloys in corrosive media like HCl, H_2SO_4, H_3PO_4, and HNO_3 [18]. Numerous naturally occurring compounds have been explored as conceivable metal CIs. The vast preponderance of these investigations has concentrated on plant extracts. The chief elements of phytoconstituents have been documented to be a diverse spectrum of organic substances such as polyphenols, terpenes, carboxylic acids, and alkaloids. As a consequence, the preponderance of these chemicals has heteroatoms (P, N, S, O atoms) and many bonds in their framework, which aid as adsorption centers for their deposition on metal surfaces. Some of these chemicals have the potential to inhibit microbial proliferation since they are remarkably efficient reactive O_2 scavengers. Polyphenol-based excerpts tend to match the bulk of the criteria for a combination of anticorrosion/antibiogradation substance, while being less poisonous than that of other plant excerpts, like alkaloid excerpts. Nevertheless, the steadiness of these extracts in the media evaluated as CIs is always a concern. Plant extracts can be prepared using a variety of methods, which can be found in the literature. The detailed discussion of these procedures is beyond the scope of this documented literature. Moreover, ILs and chemicals generated from multicomponent reactions (MCRs), ultrasonic and microwave irradiation, also have recently been developed as green methods for the inhibition of metallic corrosion. However, various factors must be considered before they may be used in real-world testing scenarios [19]. To comprehend what is green and what is not poisonousness, bioaccumulation, and biodegradability research of these compounds as mentioned above must be undertaken before testing them in the actual world. These types of tests were not carried out in the majority of the literature and publications included herein, leaving an exposed subject for future investigations. Because plant excerpts and medications are sourced from environmental and biological practices, they have a strong ability for application as environmentally friendly CIs. Numerous medications, chemical substances, and plant extracts have previously been used for a variety of engineering uses, with the nonhazardous behavior acknowledged in passing. Some examples are LD50, LC50, and EC50. Also, plant extracts of the roots, stem, seed, and other sections have been assessed, but leaf extracts have not been explored much; hence, it is endorsed that leaf extracts of these plants must be investigated [20]. It is advised that the isolation and corrosion suppression behavior of each constituent be performed out since this has been well recognized that just some particular elements of a plant extract are essential in the prevention of metallic decomposition. GC-MS and HPLC-MS procedures can be used to efficiently separate the extraction constituents (phytochemicals). Additionally, because of many significant problems in traditional separation techniques, like prolonged processing time and high operating temperature, extract treatment (extraction method) is one of the most important approaches. As a result, there is a significant requirement for eccentric extraction procedures that necessitate a short processing time and an optimal working temperature. In this regard, besides water, supercritical fluids are a novel family of different solvents for the manufacture of plant extracts that allow for the careful severance of phytochemicals from excerpts at low temperatures and short treating times. Moreover, the performance of these GCIs can still be improved. As mentioned earlier, the researchers only confirmed the usage of CIs, with intensifiers added in some instances (mainly potassium halides) [20]. Yet, if authentic corrosion inhibition interpretations were utilized, these CIs would have been far more efficacious.

Even though there have been numerous papers on the influence of halide and also other inorganic ions on the inhibitory activity of plant extracts, there is yet enough room to investigate the phenomena of synergism. The majority of the studies do not take computational methodologies into serious consideration. Recently, computer tools, particularly density functional theory (DFT)-based quantum chemical computations, have been developed. As new tools for describing the efficiency of a specific sequence of substances investigated as CIs, molecular dynamics (MD), and/or Monte Carlo (MC) simulations have appeared. Although each extract contains a range of components (phytochemicals), the additional assessment of these comparatively contemporary and greener computational tools against the metal substrate should be determined. The naiveté of these approaches stems from using software simulations that yield relatively correct estimates about the kinetics and adsorption performance of substances at material interactions without the use of hazardous substances and solvents. The MD and MC numerical simulations can deliver inhibitor (phytochemical) configurations at the contacts (horizontal or vertical). In consideration of this, it is necessary to examine the implementation of DFT and simulations (MD and MC) in the domain of metal corrosion protection utilizing natural ingredients. To summarize, phytoconstituents as metallic CIs are strongly suggested for prospective anticorrosion investigation since they are less expensive, readily accessible, compostable, biocompatible, harmless, swappable components, extremely efficient, and environmentally safe; even so, the extraction and interpretation of each phytochemical/constituent, the occurrence of synergism, the usage of phytoconstituents, and computational (DFT, MD, and MC) and outer layer assessment technologies must be addressed.

10.3 Biomass, agriculture, and biopolymer wastes and essential oils as green corrosion inhibitors: challenges and prospects

Corrosion is a worldwide problem that must be addressed correctly and quickly. When it comes to selecting a corrosion protection system, there are several factors to consider. As a result, when compared to other forms of corrosion control solutions like coatings and paint, CIs are a popular option. Conventional inhibitors, on the other hand, have been proved to be carcinogenic and environmentally hazardous. Organic CIs may therefore be employed as a substitute. Organic CIs have many qualities that contribute to metal protection, including adsorption type (physisorption, chemisorption, and comprehensive), capacity to build a protective coating on top of the metal surface, and nucleophilic and hydrophobic properties [21]. Organic GCIs manufactured from biomass waste extracts such as seed and peels, agricultural waste, biopolymers, and plant extracts such as essential oils have sparked a lot of interest in recent years. This is in line with worries about the massive volume of biomass waste and agricultural waste generated each year.

Some research gaps and future directions have been listed here based on a comprehensive literature review: Corrosion inhibition is not present in all bioactive compounds found in green extracts. As a result, it is unclear who is partial and active in this respect. The majority of the research does not look at the toxicity of green inhibitors. It has not been determined if all bioactive chemicals combine to serve as an inhibitor. Bionanomaterial has not been used in a lot of studies. Green-generated 2D material is not used in the research [21].

Bioactive chemicals may be extracted from green extracts and used independently in corrosion tests. A toxicity test may be performed before testing inhibition performance. The importance of bionanomaterial-based corrosion inhibition testing cannot be overstated. For green corrosion prevention, a greenly produced 2D material may be used.

The growing number of articles demonstrating the necessity of investigating essential oils as GCIs are obvious. In all, more than 120 publications have been published that detail the anticorrosive properties of 72 samples of essential oils isolated from 76 plant species from 18 different families [22]. Many of these articles are focused on developing CIs for iron and its alloys, which account for 79% of all papers, followed by copper and its alloys (8%), aluminum and its alloys (7%), and other metals and alloys (6%). Because of their biodegradability, eco-friendliness, cheap cost, and accessible availability, essential oils have shown a good ability to employ them as green and sustainable inhibitors against corrosion of all tested metals. The most common methodologies for evaluating the corrosion inhibition efficiency (IE) of all tested green inhibitors were weight loss and electrochemical investigations. Fourier-transform infrared spectroscopy, scanning electron microscopy (SEM) combined with energy-dispersive X-ray spectroscopy (EDX), and X-ray photoelectron spectroscopy (XPS) methods were used to examine the protective layer (film) produced on the metal surface. In general, all essential oils worked as stronger CIs of all metals in a dosage-dependent way, according to the data. With increasing concentrations of the tested natural ingredients in various test media, the IE rose considerably. This suggests that the extracts' active molecules were adsorbed onto the metal surfaces through the production of insoluble intermediates, essentially blocking the active sites and isolating the metals and corrosive agents. Furthermore, the IE of the bulk of the materials evaluated dropped as the temperature in the measured range increased. The bulk of the investigations concluded that these products function as mixed-type inhibitors based on Potentiodynamic polarization (PDP) measures. Furthermore, most CIs tested control corrosion largely by physisorption and follow the Langmuir adsorption isotherm, according to thermodynamic simulations [23]. Furthermore, essential oil's ability to suppress corrosion is strongly linked to its chemical makeup, which is made up of a variety of phytochemical components, including terpenic HCs and their oxygenated terpenoid derivatives. Because of this intricacy, determining the specific contributions of the various components to the overall inhibitory action of the extract is challenging. However, the inhibitory efficacy of these extracts might be linked to certain molecules, particularly important components or those that are prone to becoming active, or to a synergistic action of the complete extract. In this context, future studies involving interaction between individual constituents, particularly major compounds isolated from essential oils, and metal surfaces, as well as their contributions to the inhibitory effect of the entire extract will require theoretical approaches involving a computational study (DFT and Molecular dynamics simulations (MDS)).

Biopolymers/carbohydrate polymers have been investigated as potential substitutes for traditional inhibitors to solve toxicity problems. Biopolymers (i.e., tannin, lignin, extracellular polymeric material, inulin, cellulose, carrageenan, chitosan, gum, polydopamine, and others) have been found to successfully prevent alloy/metal corrosion. Biopolymers have been identified in a wide range of forms, and some of them have been tested for anticorrosion capabilities [24]. They have been subjected to a variety of alterations, and several methodologies have been used to assess their effectiveness. Despite certain limitations in the perception of biopolymers as CIs, much more effort needs to be made in terms of generating novel biopolymers and modifying current ones.

Despite the fact that significant empirical investigation has been performed and numerous research articles have been published on the issue, scientific methodology is still insufficient, culminating in subjective rather than theoretical as well as mechanistic conclusions. Instead of focusing on the specific adsorption models and related processes, the bulk of testing is done using electrochemical studies to quickly examine the effectiveness in the eventual use as corrosion preventative agents [25]. Another key issue with green extracts is the lack of identification of the active substance, which is often a mixture of undetermined chemicals. This aggravates a field where the harvesting location, weather patterns, harvesting technique, season, and a range of other elements all play a role. Procedures for sampling may be critical, resulting in contradictory findings. For example, propolis aqueous extract was used to assess corrosion resistance on carbon steel by Fouda et al. [20]. Propolis was shown to be an efficient inhibitor of carbon steel corrosion in aqueous conditions, according to the findings. The efficacy of inhibition increased when the concentration of the extract was increased, reaching up to 92%. A commercial ethanolic extract of the same substance was utilized as a CI for carbon steel, by Dolabella et al. a few years later [21]. Even though the investigated substrate (propolis) was the same and was prepared using different extraction strategies (aqueous or ethanolic or aqueous extractions), and despite the fact that both investigations reported favorable inhibitory efficacy findings, the quantitative results of the experiments were completely different. Furthermore, the inhibitory impact might be attributed to a single molecule or a synergistic effect of the complete extract. This is a feature that is seldom considered in the literature. Plant extracts, ultrasonic and microwave irradiation, and ILs and chemicals created by MCRs have recently evolved as sustainable methods for preventing metallic degradation. Nevertheless, before they can be used in real-world testing scenarios, several factors must be considered. To understand what is green and what toxicity is not bioaccumulation, biodegradability research must be carried out before testing them in the actual world [8]. These sorts of tests were not carried out in the majority of the literature and papers included here, leaving an open topic for future research. Biological and natural processes generate plant extracts and agricultural waste; thus, they have a lot of promise as CIs that are safe for the environment. Several plant extracts have been employed in a range of industrial purposes in the past, with the harmless nature mentioned very briefly (e.g., LD_{50}, LC_{50}, and EC_{50}). Because only a few specific elements of a plant extract are capable of preventing metallic deterioration, it is recommended that each ingredient be isolated and the anticorrosion action of each element be examined [8]. GC-MS and HPLC-MS methods may be used to readily isolate the extract ingredients (phytochemicals). Furthermore, because of various issues with traditional extraction processes, such as lengthy processing times and high processing temperatures, extract processing (extraction method) is one of the most essential approaches. As a result, novel extraction techniques with short processing times and optimal operating temperatures are in great demand. Moreover, typical extraction processes that necessitate the use of very toxic organic solvents have been studied, as it influences solvents used for extract production. Plant extracts, such as essential oils, are projected to be the most efficient anticorrosive agents based on the studies detailed in earlier chapters of this book. An extensive study on essential oils as well as biopolymers and agriculture and biomass waste as anticorrosion agents has been performed especially for aluminum and mild steel, and the application of these inhibitors for additional metals and alloys such as nickel, zinc, and magnesium should be investigated. Computational approaches are not taken into account in the bulk of the studies that have been published so far. Computational approaches, such as quantum studies and MD as well as

MC simulations, have recently evolved as novel techniques for describing the efficiency of CIs. Because every plant extract comprises multiple ingredients (phytochemicals), these relatively newer and greener computer tools should be used to establish their relative efficiency in relation to the surface of the metal [9]. The greenness of these approaches stems from the fact that they used software simulations to provide a relatively accurate prediction of chemical reactivity and adsorption behavior of chemical compounds at metal/electrolyte interfaces without the use of harmful chemicals or solvents.

Many studies have identified the biopolymer extraction method as a crucial step that may affect the quantity of recovered biopolymer as well as its corrosion mitigation characteristics [26-28]. The relevance of the temperature, type of electrolyte, pH, the concentration of inhibitor, and exposure period while the biopolymer is acting as an inhibitor was emphasized. In the literature, biopolymers were mostly utilized as CIs in HCl, NaCl, and saltwater; however, their amounts varied in different articles, which was claimed to impact the result. Although prior investigations employed metals such as copper and aluminum as substrates, the bulk of them was done on steel; nevertheless, the steel alloys used in each research were varied. Numerous approaches for studying inhibitor characteristics were described. In this study, many commonly used approaches for testing inhibitor characteristics were discussed. Potentiodynamic polarization, weight loss, and electrochemical impedance spectroscopy (EIS) were among the electrochemical approaches used. Other methods including SEM, EDS, EDX, and XPS were also employed. Many researches determined thermodynamic parameters (ΔH, ΔS, and E_a) [10]. The hypothesized processes involved in the inhibitor adsorption process were discussed. Although theoretical approaches are extensively used in many green inhibitor studies, they are confined to a few investigations in biopolymers, as detailed in this chapter, and further study should be done utilizing these promising methodologies [11].

10.4 Limitations of research techniques utilized in corrosion inhibition investigations

Corrosion can be monitored in a number of ways like weight loss measurement, potentiodynamic polarization measurement, and electrochemical impedance spectroscopy along with various surface morphological techniques and theoretical techniques. All these techniques are collectively utilized for calculating the corrosion inhibition efficiency and can be assumed as an inseparable part of any analysis. Weight loss is the most common, simplest, and cost-effective way to measure the effectiveness of corrosion inhibition. With the help of weight loss values, the IEs can be evaluated. Weight loss gives accurate and precise results which can be easily repeated, but the drawbacks associated with it are its destructive and time-consuming nature because long exposure of samples is required. It provides the average value of corrosion rate and sometimes sensitivity of weighing balance creates problems [29]. Further, as corrosion is an electrochemical process, electrochemical techniques are of utmost importance to study corrosion [30]. Linear polarization resistance and potentiodynamic polarization are the DC methods, and EIS and electrochemical frequency modulation are the AC methods to study the corrosion [31]. In the electrochemical techniques, one of the most frequently used is the EIS technique. The various techniques to monitor corrosion are depicted in Figure 10.2.

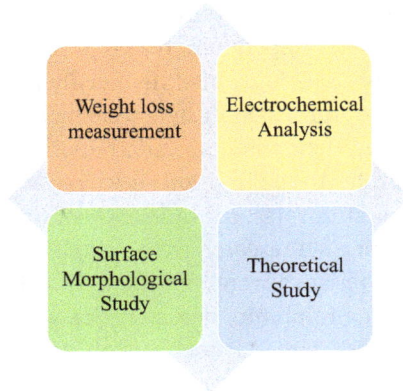

Figure 10.2: Various techniques to monitor corrosion.

EIS approach utilizes the short measurement time, comparatively inexpensive, and quantitative with high precision and repeatability of the results [32]. Generally, the equivalent circuit models were utilized to know about corrosion properties, but selecting the right circuit fit is a big challenge as in a single data more than one circuit can be fitted [33]. One more problem in these techniques is that the applied potential may cause few errors, like scan rate can alter the value of polarization resistance, specifically with a solution having low conductivity and high potential can cause the destruction of sample in potentiodynamic polarization method [34]. Potential variation can also give a false value of charge transfer resistance and rate of corrosion in EIS and electrochemical frequency modulation methods [35]. Electrochemical noise techniques also required two symmetrical electrodes throughout the reaction, which is nearly impossible because the destruction may cause during the onset of corrosion as corrosion has never been a uniform process, and other techniques like scanning vibrating electrode technique, localized EIS, and scanning electrochemical microscopy were also employed [36]. But along with benefits like high sensitivity, ease to use, and local impedance data, these techniques also come with a few disadvantages like very high time for completion, low resolution, and the effect of electrolyte conductance. Because of all these drawbacks, these methods require constant attention from researchers. Corrosion research has long relied on these established methods, but corrosion researchers still have a strong need to obtain information about corrosion in natural circumstances [37]. In most of the corrosion studies, the interpretations of electrochemical techniques are validated with surface morphological techniques like SEM, atomic force microscopy, and contact angle. The surface qualities of materials, such as texture, energy, and morphology, are rarely considered in investigations and are not given enough importance. But in reality, corrosion study does not depend upon surface energy and hydrophobicity only. Still, the texture and composition of the surface and the hydrophobic character of the surface also need in-depth analysis [38]. Specifically discussing about few surface morphological techniques like SEM, it is less time-consuming but special sample preparation is required before the time of analysis. This analysis is also costly, requires experts to handle the instruments, and can be kept in a place free from any type of electric and magnetic effects. On the other hand, atomic force microscopy is another surface morphological technique, providing three-dimensional images and even with more clarity as compared to SEM but scan rate is always an issue with this instrument. Chances of sample and tip of instrument damage always remain there [39]. One more example is the contact angle study, but the main drawbacks are that with the help of one drop, the nature of large samples cannot be accurately detected, and chances of errors are large. Conventional techniques are very time-consuming; therefore, research is now orienting toward simulation techniques to study corrosion and the mechanism involved in the process [40]. The way CIs are interacting and protect the sample can be better explained by knowing the small changes occurring on the surface of metal but a very small concentration of used inhibitor along with difficulty in understanding the reaction going on the surface, the help of theoretical study is taken to get an insight in the mechanism [40]. Moreover, these studies must be planned with great care, as incorrect methodology or a lack of data can easily lead to erroneous conclusions [41]. These theoretical studies are just the assumptions of real nature of reaction.

10.5 Conclusion

Because they are lucrative, readily accessible, decomposable, nontoxic, removable materials, highly efficient, and eco-friendly, the use of biomass, agriculture, and biopolymers waste, as well as essential oils as metallic CIs, is suggested for the upcoming corrosion inhibition exploration. Despite the increased interest, research on biomass wastes is lagging behind research on plant-based organic GCIs. Finally, it is suggested that the number of research concentrating on biomass wastes is increased. In exchange, the cost of CIs and operational costs may be decreased, and the facility can reach its full capacity while producing far less waste.

References

[1] Bashir S, Thakur A, Lgaz H, Chung IM, Kumar A. Corrosion inhibition efficiency of bronopol on aluminium in 0.5 M HCl solution: Insights from experimental and quantum chemical studies. Surfaces and Interfaces, 2020; 20: 100542. Available from: https://doi.org/10.1016/j.surfin. 2020.100542

[2] Sharma S, Ganjoo R, Saha SK, Kang N, Thakur A, Assad H, Kumar A. Investigation of inhibitive performance of Betahistine dihydrochloride on mild steel in 1M HCl solution. Journal of Molecular Liquids, 2021; 22: 118383.

[3] Bashir S, Thakur A, Lgaz H, Chung IM, Kumar A. Corrosion inhibition performance of acarbose on mild steel corrosion in acidic medium: An experimental and computational study. Arabian Journal for Science and Engineering, 2020: 45(6): 4773–4783. Available from https://doi.org/ 10.1007/s13369-020-04514-6

[4] Sharma V, Kumar S, Bashir S, Ghelichkhah Z, Obot IB, Kumar A. Use of Sapindus (reetha) as corrosion inhibitor of aluminium in acidic medium. Materials Research Express, 2018; 5(7): 076510.

[5] Sharma S, Ganjoo R, Kr. Saha S, Kang N, Thakur A, Assad H, Sharma V, Kumar A. Experimental and theoretical analysis of baclofen as a potential corrosion inhibitor for mild steel surface in HCl medium. Journal of Adhesion Science and Technology, 2021: 0(0): 1–26. Available from https://doi.org/10.1080/01694243.2021.2000230

[6] Thakur A, Kumar A. Sustainable inhibitors for corrosion mitigation in aggressive corrosive media: A comprehensive study. Journal of Bio- and Tribo-Corrosion. Springer International Publishing, 2021; 7(2): 1–48. Available from https://doi.org/10.1007/s40735-021-00501-y

[7] Assad H, Kumar A. Understanding functional group effect on corrosion inhibition efficiency of selected organic compounds. Journal of Molecular Liquids, 2021; 15(344): 117755.

[8] Lei G, Kaya C, Tüzün B, Obot IB, Touir R, Islam N. Quantum chemical and molecular dynamic simulation studies for the prediction of inhibition efficiencies of some piperidine derivatives on. Journal of the Taiwan Institute of Chemical Engineers, 2016; 0: 1–8. Available from: http://dx.doi.org/10.1016/j.jtice.2016.05.034

[9] Liu H, Gu T, Zhang G, Wang W, Dong S, Cheng Y, Liu H. Corrosion inhibition of carbon steel in CO_2-containing oilfield produced water in the presence of iron-oxidizing bacteria and inhibitors. Evaluation and Program Planning, 2016; 105: 149–160. Available from: http://dx.doi.org/10.1016/j.corsci.2016.01.012

[10] Kadhim A, Jawad RS, Numan NH. Determination the wear rate by using XRF technique for Kovar alloy under lubricated. International Journal of Computation and Applied Sciences, 2017; 2(1): 1–5.

[11] Hanoon M, Zinad DS, Resen AM, Al-Amiery AA. Gravimetrical and surface morphology studies of corrosion inhibition effects of a 4-aminoantipyrine derivative on mild steel in a corrosive solution. International Journal of Corrosion and Scale Inhibition, 2020; 9(3): 953–966.

[12] Alhaffar MT, Umoren SA, Obot IB, Ali SA. Isoxazolidine derivatives as corrosion inhibitors for low carbon steel in HCl solution: Experimental, theoretical and effect of KI studies. RSC Advances, 2018; 8(4): 1764–1777.

[13] Kaczerewska O, Leiva-Garcia R, Akid R, Brycki B, Kowalczyk I, Pospieszny T. Effectiveness of O-bridged cationic gemini surfactants as corrosion inhibitors for stainless steel in 3 M HCl: Experimental and theoretical studies. Journal of Molecular Liquids, 2018 Jan 1; 249: 1113–1124.

[14] Verma C, Quraishi MA, Ebenso EE. Microwave and ultrasound irradiations for the synthesis of environmentally sustainable corrosion inhibitors: An overview. Sustainable Chemistry and

Pharmacy, 2018; 10(November): 134–147. Available from https://doi.org/10.1016/j.scp.2018.11.001

[15] Sharma S, Kumar A. Recent advances in metallic corrosion inhibition: A review. Journal of Molecular Liquids, 2021; 322: 114862. Available from: https://doi.org/10.1016/j.molliq.2020.114862

[16] Miralrio A, Vázquez AE. Plant extracts as green corrosion inhibitors for different metal surfaces and corrosive media: A review. Processes, 2020; 8(8): 942.

[17] Gece G. Drugs: A review of promising novel corrosion inhibitors. Corrosion Science, 2011; 53 (12): 3873–3898.

[18] Marzorati S, Verotta L, Trasatti SP. Green corrosion inhibitors from natural sources and biomass wastes. Molecules, 2019; 24(1): 48.

[19] Tang Z. A review of corrosion inhibitors for rust preventative fluids. Current Opinion in Solid State & Materials Science, 2019: 23(4): 1–16. Available from https://doi.org/10.1016/j.cossms.2019.06.003

[20] Fouda AS, Badr AH. Aqueous extract of propolis as corrosion inhibitor for carbon steel in aqueous solutions. African Journal Of. Pure and Applied Chemistry, 2013; 7: 350–359.

[21] Dolabella LMP, Oliveira JG, Lins V, Matencio T, Vasconcelos WL. Ethanol extract of propolis as a protective coating for mild steel in chloride media. Journal of Coatings Technology and Research, 2016; 13: 543–555.

[22] Znini M. Application of essential oils as green corrosion inhibitors for metals and alloys in different aggressive mediums – A review. Arabian Journal of Medicinal and Aromatic Plants, 2019; 5: 3.

[23] Abdullah Dar M. A review: Plant extracts and oils as corrosion inhibitors in aggressive media. Industrial Lubrication and Tribology, 2011; 63(4): 227–233.

[24] Shahini MH, Ramezanzadeh B, Mohammadloo HE. Recent advances in biopolymers/carbohydrate polymers as effective corrosion inhibitive macro-molecules: A review study from experimental and theoretical views. Journal of Molecular Liquids, 2021; 325: 115110. Available from: https://doi.org/10.1016/j.molliq.2020.115110

[25] Fathima Sabirneeza AA, Geethanjali R, Subhashini S. Polymeric corrosion inhibitors for iron and its alloys: A review. Chemical Engineering Communications, 2015; 202(2): 232–244.

[26] Imran Ahamed M, Luqman M, Altalhi T Sustainable corrosion inhibitors Edited by. 2021. Available from: https://www.mrforum.com

[27] Yadav M, Goel G, Hatton FL, Bhagat M, Mehta SK, Mishra RK, Bhojak N. A review on biomass-derived materials and their applications as corrosion inhibitors, catalysts, food and drug delivery agents. Current Research in Green and Sustainable Chemistry, 2021; 4: 100153. Available from: https://doi.org/10.1016/j.crgsc.2021.100153

[28] Verma C, Ebenso EE, Bahadur I, Quraishi MA. An overview on plant extracts as environmental sustainable and green corrosion inhibitors for metals and alloys in aggressive corrosive media. Journal of Molecular Liquids, 2018; 266: 577–590. Available from: https://doi.org/10.1016/j.molliq.2018.06.110

[29] Bautista A, Bertocci U, Huet F. Noise resistance applied to corrosion measurements: V. Influence of electrode asymmetry. Journal of the Electrochemical Society, 2001; 148(10): B412.

[30] Fateh A, Aliofkhazraei M, Rezvanian AR. Review of corrosive environments for copper and its corrosion inhibitors. Arabian Journal of Chemistry, 2020: 13(1): 481–544. Available from https://doi.org/10.1016/j.arabjc.2017.05.021

[31] Gece G. The use of quantum chemical methods in corrosion inhibitor studies. Corrosion Science, 2008: 50(11): 2981–2992. Available from http://dx.doi.org/10.1016/j.corsci.2008.08.043

[32] Kirkland NT, Birbilis N, Staiger MP. Assessing the corrosion of biodegradable magnesium implants: A critical review of current methodologies and their limitations. Acta Biomaterialia, 2012: 8(3): 925–936. Available from http://dx.doi.org/10.1016/j.actbio.2011.11.014

[33] Dwivedi D, Lepková K, Becker T. Carbon steel corrosion: A review of key surface properties and characterization methods. RSC Advances, 2017; 7(8): 4580–4610.

[34] Obot IB, Onyeachu IB, Zeino A, Umoren SA. Electrochemical noise (EN) technique: Review of recent practical applications to corrosion electrochemistry research. Journal of Adhesion Science and Technology, 2019: 33(13): 1453–1496. Available from https://doi.org/10.1080/01694243.2019.1587224

[35] Obot IB, Macdonald DD, Gasem ZM. Density functional theory (DFT) as a powerful tool for designing new organic corrosion inhibitors: Part 1: An overview. Corrosion Science, 2015; 99 (January): 1–30. Available from http://dx.doi.org/10.1016/j.corsci.2015.01.037

[36] Sagar Dubey R, Dubey RS, Upadhyay SN. A review of electrochemical techniques applied to microbiologically influenced corrosion in recent studies. Indian Journal of Chemical Technology, 1999; 6(4): 207–218.

[37] Fajardo S, García-Galvan FR, Barranco V, Galvan JC, Batlle FS. A critical review of the application of electrochemical techniques for studying corrosion of Mg and Mg alloys: Opportunities and challenges. Magnes Alloy – Sel Issue, 2018; 1: 5-28.

[38] Jadhav N, Gelling VJ. Review – The use of localized electrochemical techniques for corrosion studies. Journal of the Electrochemical Society, 2019; 166(11): C3461–76.

[39] Feliu S, García-Galvan FR, Llorente I, Diaz L, Simancas J. Influence of hydrogen bubbles adhering to the exposed surface on the corrosion rate of magnesium alloys AZ31 and AZ61 in sodium chloride solution. Materials and Corrosion, 2017; 68(6): 651–663.

[40] Nygaard PV, Geiker MR, Elsener B. Corrosion rate of steel in concrete: Evaluation of confinement techniques for on-site corrosion rate measurements. Materials and Structures Constr., 2009; 42(8): 1059–1076.

[41] Vedalakshmi R, Thangavel K. Reliability of electrochemical techniques to predict the corrosion rate of steel in concrete structures. Arabian Journal for Science and Engineering, 2011; 36(5): 769–783.

Index

abrasion 31, 33
abrasive 19, 32, 42
accessible 192, 195–196, 201
active 141, 143, 147–148, 154, 160,
    166, 168
adsorption 61, 64–66, 67, 68, 69, 118,
    126–128, 130, 164, 166–167
adsorption isotherms 83, 87
aggressive 97, 99
agro-food wastes 80
aircraft 53
alloys and metals 75–77, 79
anodic 20, 25–30, 32–33, 37, 39, 115–119,
    121–122, 133, 140, 164
anticorrosive 98, 107
antifouling coating 155–156
anti-graffiti coatings 143, 162
anti-icing surfaces 164
antimicrobial 143–144, 156, 159–160, 162, 164,
    167–168
antismudging coatings 162
aromatic 141, 144
authors 183

bioaccumulation 191–192, 198
biodegradability 192, 198
biodegradable 156
biofilms 156
biofouling 154–155
biomass 140, 150, 161, 167
biomass waste 75, 82, 86–87
biopolymers 97–98
boiler 176

calomel 116
capacitance 120
carboxymethyl cellulose 103
carcinogenic 140
cardanol 141, 147, 151, 153–154, 156, 161, 163
Castor oil 150
catalyst 148, 150
catastrophic 20
cathodic 20, 25–30, 32, 37–40, 115–119, 122,
    133, 140, 164
cathodic inhibition 108
Cellulose 103

chalcones 141
chemisorption 51, 54
chitin 99
chitosan 97, 99, 159
chromate 140
circuit 117, 120
CNSL 147
coatings 97–108
column chromatography 141
commercial viability 167
commercialization 175
concentration 193, 198–199, 201
constituents 66–69
contact angles 152
coronavirus 159
correlation 116, 129
corrosion 1–2, 3, 4, 6–12, 19, 22–25, 29,
    33–34, 36–37, 39–44, 61–62, 64–69,
    97–99, 101–108, 115–133, 189–190, 196,
    199–200
corrosion inhibitor 75, 77, 79, 81–82, 84–85
corrosive 1–2, 7
cottonseed 145
coumarins 141
covalent 123, 129
culminating 25

dangerous 22, 34, 37–38
decomposition 147, 192
degradability 196
degradation 19, 29, 36, 39, 41, 190, 194, 198
descaling 1
destruction 115, 119, 123, 133
destructive 57
deterioration 19, 21–23, 26, 37, 39–40, 43, 97,
    105, 175, 185, 191, 198
DFT 165–166
disintegration 19, 23, 26, 41

eco-friendly 97, 104–105
economic 52–53, 176, 180
economy 53
EDX 116, 123–124
efficiency 2–3, 7, 10–12, 62–66, 68, 140, 143,
    145, 150–151, 154–155, 160–161, 166, 169,
    190, 195, 199

https://doi.org/10.1515/9783110760583-011

www.ingramcontent.com/pod-product-compliance
Lightning Source LLC
Chambersburg PA
CBHW061416210326
41598CB00035B/6229